CW00501433

SOURCES OF INFORMATION ON MOTION

> " No place affords a more striking conviction of
> the vanity of human hopes than a public library. "
> Samuel Johnson

> " In a consumer society there are inevitably two
> kinds of slaves: the prisoners of addiction and
> the prisoners of envy. "
> Ivan Illich**

In the text, good books that introduce neighbouring domains are presented in the bibliography. The bibliography also points to journals and websites, in order to satisfy more intense curiosity about what is encountered in this adventure. All citations can also be found by looking up the author in the name index. To find additional information, either libraries or the internet can help.

In a library, review articles of recent research appear in journals such as Reviews of Modern Physics, Reports on Progress in Physics, Contemporary Physics and Advances in Physics. Good pedagogical introductions are found in the American Journal of Physics, the European Journal of Physics and Physik in unserer Zeit.

Overviews on research trends occasionally appear in magazines such as Physics World, Physics Today, Europhysics Journal, Physik Journal and Nederlands tijdschrift voor natuurkunde. For coverage of all the sciences together, the best sources are the magazines Nature, New Scientist, Naturwissenschaften, La Recherche and Science News.

Research papers on the foundations of motion appear mainly in Physics Letters B, Nuclear Physics B, Physical Review D, Physical Review Letters, Classical and Quantum Gravity, General Relativity and Gravitation, International Journal of Modern Physics and Modern Physics Letters. The newest results and speculative ideas are found in conference proceedings, such as the Nuclear Physics B Supplements. Research articles also appear in Fortschritte der Physik, European Physical Journal, La Rivista del Nuovo Cimento, Europhysics Letters, Communications in Mathematical Physics, Journal of Mathematical Physics, Foundations of Physics, International Journal of Theoretical Physics and Journal of Physics G.

There are only a few internet physics journals of quality: one is Living Reviews in Relativity, found at www.livingreviews.org, the other is the New Journal of Physics, which can be found at the www.njp.org website. There are, unfortunately, also many internet physics journals that publish incorrect research. They are easy to spot: they ask for money

---

** Ivan Illich (b. 1926 Vienna, d. 2002 Bremen), theologian and social and political thinker.

to publish a paper.

By far the simplest way to keep in touch with ongoing research on motion and modern physics is to use the *internet*, the international computer network. To start using it, ask a friend who knows.*

In the last decade of the twentieth century, the internet expanded into a combination of library, business tool, discussion platform, media collection, garbage collection and, above all, addiction provider. Do not use it too much. Commerce, advertising and – unfortunately – addictive material for children, youth and adults, as well as crime of all kind are also an integral part of the web. With a personal computer, a modem and free browser software, you can look for information in millions of pages of documents or destroy your professional career through addiction. The various parts of the documents are located in various computers around the world, but the user does not need to be aware of this.**

Most theoretical physics papers are available free of charge, as *preprints*, i.e., before official publication and checking by referees, at the arxiv.org website. A service for finding subsequent preprints that cite a given one is also available.

Research papers on the description of motion appear *after* this text is published can also be found via www.webofknowledge.com a site accessible only from libraries. It allows one to search for all publications which *cite* a given paper.

Searching the web for authors, organizations, books, publications, companies or simple keywords using search engines can be a rewarding experience or an episode of addiction, depending entirely on yourself. A selection of interesting servers about motion is given below.

---

* It is also possible to use the internet and to download files through FTP with the help of email only. But the tools change too often to give a stable guide here. Ask your friend.

** Several decades ago, the provocative book by IVAN ILLICH, *Deschooling Society*, Harper & Row, 1971, listed four basic ingredients for any educational system:

1. access to *resources* for learning, e.g. books, equipment, games, etc. at an affordable price, for everybody, at any time in their life;

2. for all who want to learn, access to *peers* in the same learning situation, for discussion, comparison, cooperation and competition;

3. access to *elders*, e.g. teachers, for their care and criticism towards those who are learning;

4. exchanges between students and *performers* in the field of interest, so that the latter can be models for the former. For example, there should be the possibility to listen to professional musicians and reading the works of specialist writers. This also gives performers the possibility to share, advertise and use their skills.

Illich develops the idea that if such a system were informal – he then calls it a 'learning web' or 'opportunity web' – it would be superior to formal, state-financed institutions, such as conventional schools, for the development of mature human beings. These ideas are deepened in his following works, *Deschooling Our Lives*, Penguin, 1976, and *Tools for Conviviality*, Penguin, 1973.

Today, any networked computer offers *email* (electronic mail), FTP (file transfers to and from another computer), access to discussion groups on specific topics, such as particle physics, and the *world-wide web*. In a rather unexpected way, all these facilities of the internet have transformed it into the backbone of the 'opportunity web' discussed by Illich. However, as in any school, it strongly depends on the user's discipline whether the internet actually does provide a learning web or an entry into addiction.

TABLE 61 Some interesting sites on the world-wide web.

Topic	Website address
General knowledge	
Innovation in science and technology	www.innovations-report.de
Book collections	www.ulib.org
	books.google.com
Entertaining science education by Theodore Gray	www.popsci.com/category/popsci-authors/theodore-gray
Entertaining and professional science education by Robert Krampf	thehappyscientist.com
Science Frontiers	www.science-frontiers.com
Science Daily News	www.sciencedaily.com
Science News	www.sciencenews.org
Encyclopedia of Science	www.daviddarling.info
Interesting science research	www.max-wissen.de
Quality science videos	www.vega.org.uk
ASAP Science videos	plus.google.com/101786231119207015313/posts
Physics	
Learning physics with toys from rubbish	www.arvindguptatoys.com
Official SI unit website	www.bipm.fr
Unit conversion	www.chemie.fu-berlin.de/chemistry/general/units.html
Particle data	pdg.web.cern.ch
Engineering data and formulae	www.efunda.com
Information on relativity	math.ucr.edu/home/baez/relativity.html
Research preprints	arxiv.org
	www.slac.stanford.edu/spires
Abstracts of papers in physics journals	www.osti.gov
Many physics research papers	sci-hub.tv, sci-hub.la
	libgen.pw, libgen.io
Physics news, weekly	www.aip.org/physnews/update
Physics news, daily	phys.org
Physics problems by Yacov KantorKantor, Yacov	www.tau.ac.il/~kantor/QUIZ/
Physics problems by Henry Greenside	www.phy.duke.edu/~hsg/physics-challenges/challenges.html
Physics 'question of the week'	www.physics.umd.edu/lecdem/outreach/QOTW/active
Physics 'miniproblem'	www.nyteknik.se/miniproblemet
Physikhexe	physik-verstehen-mit-herz-und-hand.de/html/de-6.html

TOPIC	WEBSITE ADDRESS
Magic science tricks	www.sciencetrix.com
Physics stack exchange	physics.stackexchange.com
'Ask the experts'	www.sciam.com/askexpert_directory.cfm
Nobel Prize winners	www.nobel.se/physics/laureates
Videos of Nobel Prize winner talks	www.mediatheque.lindau-nobel.org
Pictures of physicists	www.if.ufrj.br/famous/physlist.html
Physics organizations	www.cern.ch
	www.hep.net
	www.nikhef.nl
	www.het.brown.edu/physics/review/index.html
Physics textbooks on the web	www.physics.irfu.se/CED/Book
	www.biophysics.org/education/resources.htm
	www.lightandmatter.com
	www.physikdidaktik.uni-karlsruhe.de/index_en.html
	www.feynmanlectures.info
	hyperphysics.phy-astr.gsu.edu/hbase/hph.html
	www.motionmountain.net
Three beautiful French sets of notes on classical mechanics and particle theory	feynman.phy.ulaval.ca/marleau/notesdecours.htm
The excellent *Radical Freshman Physics* by David Raymond	www.physics.nmt.edu/~raymond/teaching.html
Physics course scripts from MIT	ocw.mit.edu/courses/physics/
Physics lecture scripts in German and English	www.akleon.de
'World lecture hall'	wlh.webhost.utexas.edu
Optics picture of the day	www.atoptics.co.uk/opod.htm
Living Reviews in Relativity	www.livingreviews.org
Wissenschaft in die Schulen	www.wissenschaft-schulen.de
Videos of Walter Lewin'sIndexLewin, Walter physics lectures	ocw.mit.edu/courses/physics/ 8-01-physics-i-classical-mechanics-fall-1999/
Physics videos of Matt Carlson	www.youtube.com/sciencetheater
Physics videos by the University of Nottingham	www.sixtysymbols.com
Physics lecture videos	www.coursera.org/courses?search=physics
	www.edx.org/course-list/allschools/physics/allcourses

Mathematics

TOPIC	WEBSITE ADDRESS
'Math forum' internet resource collection	mathforum.org/library
Biographies of mathematicians	www-history.mcs.st-andrews.ac.uk/BiogIndex.html
Purdue math problem of the week	www.math.purdue.edu/academics/pow
Macalester College maths problem of the week	mathforum.org/wagon
Mathematical formulae	dlmf.nist.gov
Weisstein's World of Mathematics	mathworld.wolfram.com
Functions	functions.wolfram.com
Symbolic integration	www.integrals.com
Algebraic surfaces	www.mathematik.uni-kl.de/{~}hunt/drawings.html
Math lecture videos, in German	www.j3l7h.de/videos.html
Gazeta Matematica, in Romanian	www.gazetamatematica.net

Astronomy

ESA	sci.esa.int
NASA	www.nasa.gov
Hubble space telescope	hubble.nasa.gov
Sloan Digital Sky Survey	skyserver.sdss.org
The 'cosmic mirror'	www.astro.uni-bonn.de/~dfischer/mirror
Solar System simulator	space.jpl.nasa.gov
Observable satellites	liftoff.msfc.nasa.gov/RealTime/JPass/20
Astronomy picture of the day	antwrp.gsfc.nasa.gov/apod/astropix.html
The Earth from space	www.visibleearth.nasa.gov
From Stargazers to Starships	www.phy6.org/stargaze/Sintro.htm
Current solar data	www.n3kl.org/sun

Specific topics

Sonic wonders to visit in the world	www.sonicwonders.org
Encyclopedia of photonics	www.rp-photonics.com
Chemistry textbook, online	chemed.chem.wisc.edu/chempaths/GenChem-Textbook
Minerals	webmineral.com
	www.mindat.org
Geological Maps	onegeology.org
Optical illusions	www.sandlotscience.com
Rock geology	sandatlas.org
Petit's science comics	www.jp-petit.org
Physical toys	www.e20.physik.tu-muenchen.de/~cucke/toylinke.htm

TOPIC	WEBSITE ADDRESS
Physics humour	www.dctech.com/physics/humor/biglist.php
Literature on magic	www.faqs.org/faqs/magic-faq/part2
music library, searchable by tune	imslp.org
Making paper aeroplanes	www.pchelp.net/paper_ac.htm
	www.ivic.qc.ca/~aleexpert/aluniversite/klinevogelmann.html
Small flying helicopters	pixelito.reference.be
Science curiosities	www.wundersamessammelsurium.info
Ten thousand year clock	www.longnow.org
Gesellschaft Deutscher Naturforscher und Ärzte	www.gdnae.de
Pseudoscience	suhep.phy.syr.edu/courses/modules/PSEUDO/pseudo_main.html
Crackpots	www.crank.net
Periodic table with videos for each element	www.periodicvideos.com
Mathematical quotations	math.furman.edu/mwoodard/~mquot.html
The 'World Question Center'	www.edge.org/questioncenter.html
Plagiarism	www.plagiarized.com
Hoaxes	www.museumofhoaxes.com
Encyclopedia of Earth	www.eoearth.org
This is colossal	thisiscolossal.com

Do you want to study physics without actually going to university? Nowadays it is possible to do so via email and internet, in German, at the University of Kaiserslautern.* In the near future, a nationwide project in Britain should allow the same for English-speaking students. As an introduction, use the latest update of this physics text!

> Das Internet ist die offenste Form der geschlossenen Anstalt.**
> Matthias Deutschmann

> Si tacuisses, philosophus mansisses.***
> After Boethius.

* See the www.fernstudium-physik.de website.
** 'The internet is the most open form of a closed institution.'
*** 'If you had kept quiet, you would have remained a philosopher.' After the story Boethius (c. 480–c. 525) tells in *De consolatione philosophiae*, 2.7, 67 ff.

CHALLENGE HINTS AND SOLUTIONS

“ Never make a calculation before you know the
answer. ”
John Wheeler's motto

John Wheeler wanted people to estimate, to try and to guess; but not saying the guess out loud. A correct guess reinforces the physics instinct, whereas a wrong one leads to the pleasure of surprise. Guessing is thus an important first step in solving every problem.

Teachers have other criteria to keep in mind. Good problems can be solved on different levels of difficulty, can be solved with words or with images or with formulae, activate knowledge, concern real world applications, and are open.

Challenge 1, page 10: Do not hesitate to be demanding and strict. The next edition of the text will benefit from it.

Challenge 2, page 16: There are many ways to distinguish real motion from an illusion of motion: for example, only real motion can be used to set something else into motion. In addition, the motion illusions of the figures show an important failure; nothing moves if the head and the paper remain fixed with respect to each other. In other words, the illusion only *amplifies* existing motion, it does not *create* motion from nothing.

Challenge 3, page 17: Without detailed and precise experiments, both sides can find examples to prove their point. Creation is supported by the appearance of mould or bacteria in a glass of water; creation is also supported by its opposite, namely traceless disappearance, such as the disappearance of motion. However, conservation is supported and creation falsified by all those investigations that explore assumed cases of appearance or disappearance in full detail.

Challenge 4, page 19: The amount of water depends on the shape of the bucket. The system chooses the option (tilt or straight) for which the centre of gravity is lowest.

Challenge 5, page 20: To simplify things, assume a cylindrical bucket. If you need help, do the experiment at home. For the reel, the image is misleading: the rim on which the reel advances has a *larger* diameter than the section on which the string is wound up. The wound up string does not touch the floor, like for the reel shown in Figure 304.

Challenge 6, page 19: Political parties, sects, helping organizations and therapists of all kinds are typical for this behaviour.

Challenge 7, page 24: The issue is not yet completely settled for the motion of empty space, such as in the case of gravitational waves. Thus, the motion of empty space might be an exception. In any case, empty space is not made of small particles of finite size, as this would contradict the transversality of gravity waves.

Challenge 8, page 26: Holes are not physical systems, because in general they cannot be tracked.

Challenge 9, page 26: The circular definition is: objects are defined as what moves with respect

FIGURE 304 The assumed shape for the reel puzzle.

FIGURE 305
A soap
bubble while
bursting
(© Peter
Wienerr-
oither).

to the background, and the background is defined as what stays when objects change. We shall return to this important issue several times in our adventure. It will require a certain amount of patience to solve it, though.

Page 437

Challenge 10, page 28: No, the universe does not have a state. It is not measurable, not even in principle. See the discussion on the issue in volume IV, on quantum theory.

Vol. IV, page 169

Challenge 11, page 28: The final list of intrinsic properties for physical systems found in nature is given in volume V, in the section of particle physics. And of course, the universe has no intrinsic, permanent properties. None of them are measurable for the universe as a whole, not even in principle.

Vol. V, page 262

Challenge 12, page 31: Hint: yes, there is such a point.

Challenge 13, page 31: See Figure 305 for an intermediate step. A bubble bursts at a point, and then the rim of the hole increases rapidly, until it disappears on the antipodes. During that process the remaining of the bubble keeps its spherical shape, as shown in the figure. For a film of the process, see www.youtube.com/watch?v=dIZwQ24_OU0 (or search for 'bursting soap bubble'). In other words, the final droplets that are ejected stem from the point of the bubble which is opposite to the point of puncture; they are never ejected from the centre of the bubble.

Vol. IV, page 136

Challenge 14, page 31: A ghost can be a moving image; it cannot be a moving object, as objects cannot interpenetrate.

Challenge 15, page 31: If something could stop moving, motion could disappear into nothing. For a precise proof, one would have to show that no atom moves any more. So far, this has never been observed: motion is conserved. (Nothing in nature can disappear into nothing.)

Challenge 16, page 31: This would indeed mean that space is infinite; however, it is impossible to observe that something moves 'forever': nobody lives that long. In short, there is no way to prove that space is infinite in this way. In fact, there is no way to prove that space if infinite in any other way either.

Challenge 17, page 31: The necessary rope length is nh, where n is the number of wheels/pulleys. And yes, the farmer is indeed doing something sensible.

Challenge 19, page 31: How would you measure this?

Challenge 20, page 31: The number of reliable digits of a measurement result is a simple quantification of precision. More details can be found by looking up 'standard deviation' in the index.

Challenge 21, page 31: No; memory is needed for observation and measurements. This is the case for humans and measurement apparatus. Quantum theory will make this particularly clear.

Challenge 22, page 31: Note that you never have observed zero speed. There is always some measurement error which prevents one to say that something is zero. No exceptions!

Challenge 23, page 32: $(2^{64} - 1) = 18\,446\,744\,073\,700\,551\,615$ grains of wheat, with a grain weight of 40 mg, are 738 thousand million tons. Given a world harvest in 2006 of 606 million tons, the grains amount to about 1200 years of the world's wheat harvests.

The grain number calculation is simplified by using the formula $1 + m + m^2 + m^3 + ...m^n = (m^{n+1} - 1)/(m - 1)$, that gives the sum of the so-called *geometric sequence*. The name is historical and is used as a contrast to the *arithmetic sequence* $1 + 2 + 3 + 4 + 5 + ...n = n(n + 1)/2$. Can you prove the two expressions?

The chess legend is mentioned first by Ibn Khallikan (b. 1211 Arbil, d. 1282 Damascus). King Shiram and king Balhait, also mentioned in the legend, are historical figures that lived between the second and fourth century CE. The legend appears to have combined two different stories. Indeed, the calculation of grains appears already in the year 947, in the famous text *Meadows of Gold and Mines of Precious Stones* by Al-Masudi (b. *c.* 896 Baghdad, d. 956 Cairo).

Challenge 24, page 32: In clean experiments, the flame leans forward. But such experiments are not easy, and sometimes the flame leans backward. Just try it. Can you explain both observations?

Challenge 25, page 32: Accelerometers are the simplest motion detectors. They exist in form of piezoelectric devices that produce a signal whenever the box is accelerated and can cost as little as one euro. Another accelerometer that might have a future is an interference accelerometer that makes use of the motion of an interference grating; this device might be integrated in silicon. Other, more precise accelerometers use gyroscopes or laser beams running in circles.

Velocimeters and position detectors can also detect motion; they need a wheel or at least an optical way to look out of the box. Tachographs in cars are examples of velocimeters, computer mice are examples of position detectors.

A cheap enough device would be perfect to measure the speed of skiers or skaters. No such device exists yet.

Challenge 26, page 32: The ball rolls (or slides) towards the centre of the table, as the table centre is somewhat nearer to the centre of the Earth than the border; then the ball shoots over, performing an oscillation around the table centre. The period is 84 min, as shown in challenge 405. (This has never been observed, so far. Why?)

Challenge 27, page 32: Only if the acceleration never vanishes. Accelerations can be felt. Accelerometers are devices that measure accelerations and then deduce the position. They are used in aeroplanes when flying over the atlantic. If the box does not accelerate, it is impossible to say whether it moves or sits still. It is even impossible to say in which direction one moves. (Close your eyes in a train at night to confirm this.)

Challenge 28, page 32: The block moves twice as fast as the cylinders, independently of their radius.

Challenge 29, page 32: This methods is known to work with other fears as well.

Challenge 30, page 33: Three couples require 11 passages. Two couples require 5. For four or more couples there is no solution. What is the solution if there are n couples and $n - 1$ places on the boat?

Challenge 31, page 33: Hint: there is an infinite number of such shapes. These curves are called also *Reuleaux curves*. Another hint: The 20 p and 50 p coins in the UK have such shapes. And yes, other shapes than cylinders are also possible: take a twisted square bar, for example.

Challenge 32, page 33: If you do not know, ask your favourite restorer of old furniture.

Challenge 33, page 33: For this beautiful puzzle, see arxiv.org/abs/1203.3602.

Challenge 34, page 33: Conservation, relativity and minimization are valid generally. In some rare processes in nuclear physics, motion invariance (reversibility) is broken, as is mirror invariance. Continuity is known not to be valid at smallest length and time intervals, but no experiments has yet probed those domains, so that it is still valid in practice.

Challenge 35, page 34: In everyday life, this is correct; what happens when quantum effects are taken into account?

Challenge 36, page 36: Take the average distance change of two neighbouring atoms in a piece of quartz over the last million years. Do you know something still slower?

Challenge 37, page 37: There is only one way: compare the velocity to be measured with the speed of light – using cleverly placed mirrors. In fact, almost all physics textbooks, both for schools and for university, start with the definition of space and time. Otherwise excellent relativity textbooks have difficulties avoiding this habit, even those that introduce the now standard k-calculus (which is in fact the approach mentioned here). Starting with speed is the most logical and elegant approach. But it is possible to compare speeds without metre sticks and clocks. Can you devise a method?

Challenge 38, page 37: There is no way to sense your own motion if you are in a vacuum. No way in principle. This result is often called the *principle of relativity*.

Page 156

In fact, there is a way to measure your motion in space (though not in vacuum): measure your speed with respect to the cosmic background radiation. So we have to be careful about what is implied by the question.

Challenge 39, page 37: The wing load W/A, the ratio between weight W and wing area A, is obviously proportional to the third root of the weight. (Indeed, $W \sim l^3$, $A \sim l^2$, l being the dimension of the flying object.) This relation gives the green trend line.

The wing load W/A, the ratio between weight W and wing area A, is, like all forces in fluids, proportional to the square of the cruise speed v: we have $W/A = v^2 0.38\,\text{kg/m}^3$. The unexplained

FIGURE 306 Sunbeams in a forest
(© Fritz Bieri and Heinz Rieder).

factor contains the density of air and a general numerical coefficient that is difficult to calculate. This relation connects the upper and lower horizontal scales in the graph.

As a result, the cruise speed scales as the *sixth root* of weight: $v \sim W^{1/6}$. In other words, an Airbus A380 is 750 000 million times heavier than a fruit fly, but only a hundred times as fast.

Challenge 41, page 41: Equivalently: do points in space exist? The final part of our adventure
Vol. VI, page 65
explores this issue in detail.

Challenge 42, page 42: All electricity sources must use the same phase when they feed electric power into the net. Clocks of computers on the internet must be synchronized.

Challenge 43, page 42: Note that the shift increases quadratically with time, not linearly.

Challenge 44, page 43: Galileo measured time with a scale (and with other methods). His stopwatch was a water tube that he kept closed with his thumb, pointing into a bucket. To start the stopwatch, he removed his thumb, to stop it, he put it back on. The volume of water in the bucket then gave him a measure of the time interval. This is told in his famous book GALILEO GALILEI, *Discorsi e dimostrazioni matematiche intorno a due nuove scienze attenenti alla mecanica e i movimenti locali*, usually simply called the 'Discorsi', which he published in 1638 with Louis Elsevier in Leiden, in the Netherlands.

Challenge 45, page 44: Natural time is measured with natural motion. Natural motion is the motion of light. Natural time is thus defined with the motion of light.

Challenge 46, page 48: There is no way to define a local time at the poles that is consistent with all neighbouring points. (For curious people, check the website www.arctic.noaa.gov/gallery_np. html.)

Challenge 48, page 50: The forest is full of light and thus of light rays: they are straight, as shown by the sunbeams in Figure 306.

Challenge 49, page 50: One pair of muscles moves the lens along the third axis by deforming the eye from prolate to spherical to oblate.

Challenge 50, page 50: You can solve this problem by trying to think in four dimensions. (Train using the well-known three-dimensional projections of four-dimensional cubes.) Try to imagine how to switch the sequence when two pieces cross. Note: it is usually *not* correct, in this domain, to use time instead of a fourth *spatial* dimension!

Challenge 51, page 52: Measure distances using light.

Challenge 54, page 56: It is easier to work with the unit torus. Take the unit interval $[0, 1]$ and equate the end points. Define a set B in which the elements are a given real number b from the interval plus all those numbers who differ from that real by a rational number. The unit circle can be thought as the union of all the sets B. (In fact, every set B is a shifted copy of the rational numbers $\mathbb{Q}$.) Now build a set A by taking one element from each set B. Then build the set family consisting of the set A and its copies A_q shifted by a rational q. The union of all these sets is the unit torus. The set family is countably infinite. Then divide it into *two* countably infinite set families. It is easy to see that each of the two families can be renumbered and its elements shifted in such a way that each of the two families forms a unit torus.

Mathematicians say that there is no countably infinitely additive measure of $\mathbb{R}^n$ or that sets such as A are non-measurable. As a result of their existence, the 'multiplication' of lengths is possible. Later on we shall explore whether bread or *gold* can be multiplied in this way.

Ref. 44

Challenge 55, page 56: Hint: start with triangles.

Challenge 56, page 56: An example is the region between the x-axis and the function which assigns 1 to every transcendental and 0 to every non-transcendental number.

Challenge 57, page 57: We use the definition of the function of the text. The dihedral angle of a regular tetrahedron is an irrational multiple of π, so the tetrahedron has a non-vanishing Dehn invariant. The cube has a dihedral angle of $\pi/2$, so the Dehn invariant of the cube is 0. Therefore, the cube is not equidecomposable with the regular tetrahedron.

Challenge 58, page 58: If you think you can show that empty space is continuous, you are wrong. Check your arguments. If you think you can prove the opposite, you *might* be right – but only if you already know what is explained in the final part of the text. If that is not the case, check your arguments. In fact, time is neither discrete nor continuous.

Challenge 60, page 59: Obviously, we use light to check that the plumb line is straight, so the two definitions must be the same. This is the case because the field lines of gravity are also possible paths for the motion of light. However, this is not always the case; can you spot the exceptions?

Another way to check straightness is along the surface of calm water.

A third, less precise, way is to make use of the straightness sensors on the brain. The human brain has a built-in faculty to determine whether an objects seen with the eyes is straight. There are special cells in the brain that fire when this is the case. Any book on vision perception tells more about this topic.

Challenge 61, page 60: The hollow Earth theory is correct if the distance formula is used consistently. In particular, one has to make the assumption that objects get smaller as they approach the centre of the hollow sphere. Good explanations of all events are found on www.geocities.com/inversedearth. Quite some material can be found on the internet, also under the names of celestrocentric system, inner world theory or concave Earth theory. There is no way to prefer one description over the other, except possibly for reasons of simplicity or intellectual laziness.

Challenge 63, page 61: A hint is given in Figure 307. For the measurement of the speed of light with almost the same method, see volume II, on page 20.

Challenge 64, page 61: A fast motorbike is faster: a motorbike driver can catch an arrow, a stunt that was shown on the German television show 'Wetten dass' in the year 2001.

Challenge 65, page 61: The 'only' shape that prevents a cover to fall into the hole beneath is a circular shape. Actually, slight deviations from the circular shape are also allowed.

Challenge 68, page 61: The walking speed of older men depends on their health. If people walk faster than 1.4 m/s, they are healthy. The study concluded that the grim reaper walks with a preferred speed of 0.82 m/s, and with a maximum speed of 1.36 m/s.

Ref. 378

Challenge 69, page 61: 72 stairs.

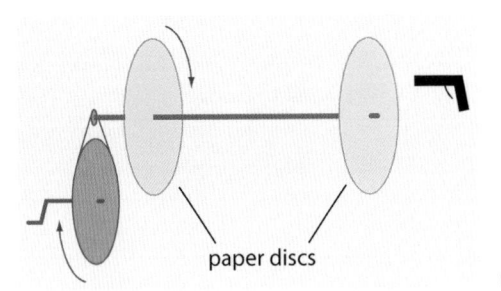

FIGURE 307 A simple way to measure bullet speeds.

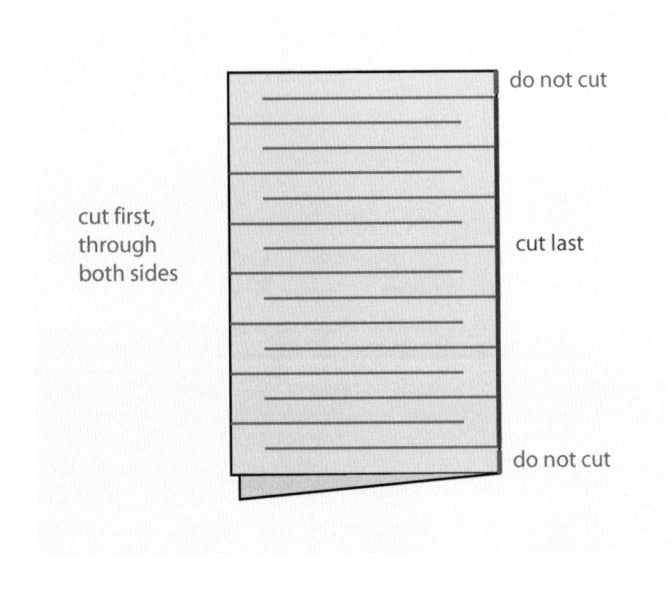

FIGURE 308 How to make a hole in a postcard that allows stepping through it.

Challenge 73, page 62: See Figure 308 for a way to realize the feat.

Challenge 74, page 62: Within 1 per cent, one *fifth* of the height must be empty, and four fifths must be filled; the exact value follows from $\sqrt[3]{2} = 1.25992...$

Challenge 75, page 62: One pencil draws a line of between 20 and 80 km, if no lead is lost when sharpening. Numbers for the newly invented plastic, flexible pencils are not available.

Challenge 79, page 63: The bear is white, because the obvious spot of the house is at the North pole. But there are infinitely many additional spots (without bears) near the South pole: can you find them?

Challenge 80, page 63: We call L the initial length of the rubber band, v the speed of the snail relative to the band and V the speed of the horse relative to the floor. The speed of the snail relative to the floor is given as

$$\frac{ds}{dt} = v + V\frac{s}{L + Vt} \,. \tag{128}$$

This is a so-called *differential equation* for the unknown snail position $s(t)$. You can check – by simple insertion – that its solution is given by

$$s(t) = \frac{v}{V}(L + Vt)\ln(1 + Vt/L) \,. \tag{129}$$

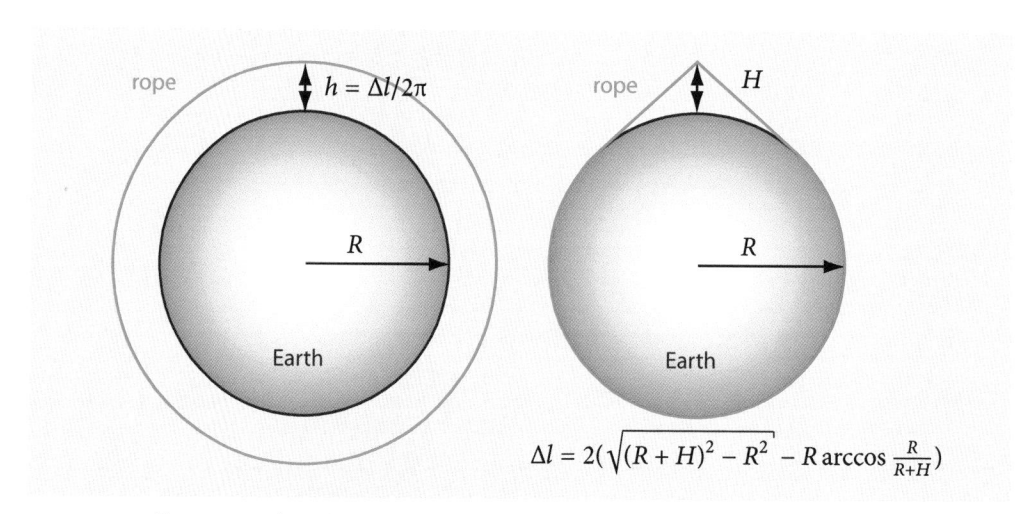

$$\Delta l = 2(\sqrt{(R+H)^2 - R^2} - R \arccos \tfrac{R}{R+H})$$

FIGURE 309 Two ways to lengthen a rope around the Earth.

Therefore, the snail reaches the horse at a time

$$t_{\text{reaching}} = \frac{L}{V}(e^{V/v} - 1) \qquad\qquad (130)$$

which is finite for all values of L, V and v. You can check however, that the time is very large indeed, if realistic speed values are used.

Challenge 81, page 63: Colour is a property that applies only to objects, not to boundaries. In the mentioned case, only spots and backgrounds have colours. The question shows that it is easy to ask questions that make no sense also in physics.

Challenge 82, page 63: You can do this easily yourself. You can even find websites on the topic.

Challenge 84, page 63: Clocks with two hands: 22 times. Clocks with three hands: 2 times.

Challenge 85, page 64: 44 times.

Challenge 86, page 64: For two hands, the answer is 143 times.

Challenge 87, page 64: The Earth rotates with 15 minutes per minute.

Challenge 88, page 64: You might be astonished, but no reliable data exist on this question. The highest speed of a throw measured so far seems to be a 45 m/s cricket bowl. By the way, much more data are available for speeds achieved with the help of rackets. The c. 70 m/s of fast badminton smashes seem to be a good candidate for record racket speed; similar speeds are achieved by golf balls.

Challenge 89, page 64: A *spread out* lengthening by 1 m allows even many cats to slip through, as shown on the left side of Figure 309. But the right side of the figure shows a better way to use the extra rope length, as Dimitri Yatsenko points out: a *localized* lengthening by 1 mm then already yields a height of 1.25 m, allowing a child to walk through. In fact, a lengthening by 1 m performed in this way yields a peak height of 121 m!

Challenge 90, page 64: 1.8 km/h or 0.5 m/s.

Challenge 92, page 64: The question makes sense, especially if we put our situation in relation to the outside world, such as our own family history or the history of the universe. The different usage reflects the idea that we are able to determine our position by ourselves, but not the time in which we are. The section on determinism will show how wrong this distinction is.

Page 238

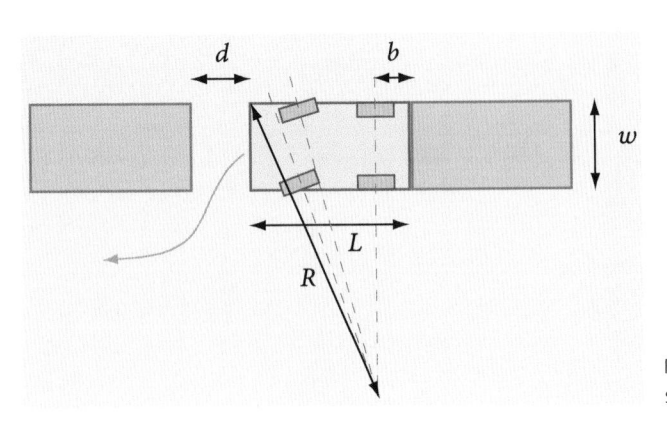

FIGURE 310 Leaving a parking space – the outer turning radius.

Challenge 93, page 64: Yes, there is. However, this is not obvious, as it implies that space and time are not continuous, in contrast to what we learn in primary school. The answer will be found in the final part of this text.

Challenge 94, page 64: For a curve, use, at each point, the curvature radius of the circle approximating the curve in that point; for a surface, define two directions in each point and use two such circles along these directions.

Challenge 95, page 64: It moves about 1 cm in 50 ms.

Challenge 96, page 65: The surface area of the lung is between 100 and 200 m^2, depending on the literature source, and that of the intestines is between 200 and 400 m^2.

Challenge 97, page 65: A limit does not exist in classical physics; however, there is one in nature which appears as soon as quantum effects are taken into account.

Challenge 98, page 65: The final shape is a full cube without any hole.

Challenge 99, page 65: The required gap d is

$$d = \sqrt{(L-b)^2 - w^2 + 2w\sqrt{R^2 - (L-b)^2}} - L + b \, , \qquad (131)$$

as deduced from Figure 310. See also R. HOYLE, *Requirements for a perfect s-shaped parallel parking maneuvre in a simple mathematical model*, 2003. In fact, the mathematics of parallel parking is beautiful and interesting. See, for example, the web page rigtriv.wordpress.com/2007/10/01/parallel-parking/ or the explanation in EDWARD NELSON, *Tensor Analysis*, Princeton University Press, 1967, pp. 33–36. Nelson explains how to define vector fields that change the four-dimensional configuration of a car, and how to use their algebra to show that a car can leave parking spaces with arbitrarily short distances to the cars in front and in the back.

Challenge 100, page 65: A smallest gap does not exist: any value will do! Can you show this?

Challenge 101, page 66: The following solution was proposed by Daniel Hawkins.

Assume you are sitting in car A, parked behind car B, as shown in Figure 311. There are two basic methods for exiting a parking space that requires the reverse gear: rotating the car to move the centre of rotation away from (to the right of) car B, and shifting the car downward to move the centre of rotation away from (farther below) car B. The first method requires car A to be partially diagonal, which means that the method will not work for d less than a certain value, essentially the value given above, when no reverse gear is needed. We will concern ourselves with the second method (pictured), which will work for an infinitesimal d.

In the case where the distance d is less than the minimum required distance to turn out of the parking space without using the reverse gear for a given geometry L, w, b, R, an attempt to

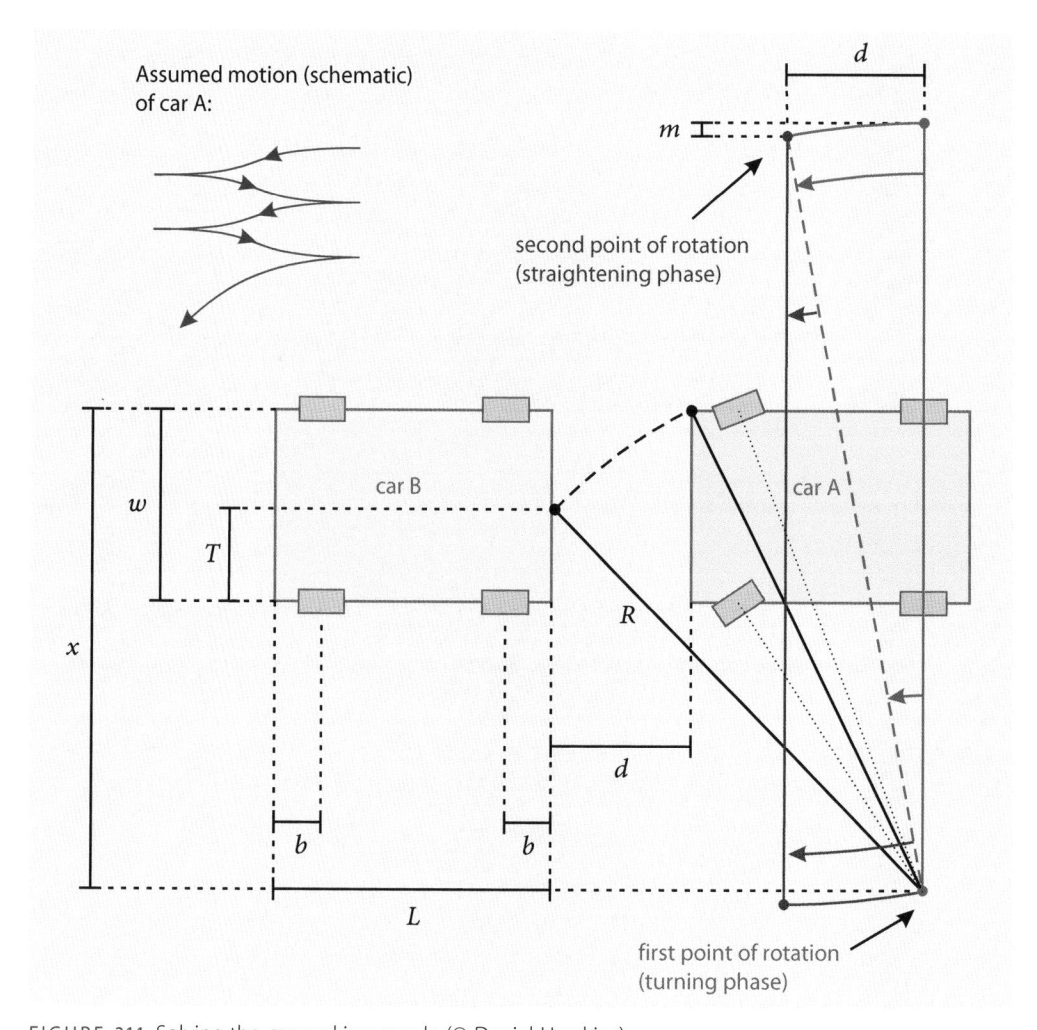

FIGURE 311 Solving the car parking puzzle (© Daniel Hawkins).

turn out of the parking space will result in the corner of car A touching car B at a distance T away from the edge of car B, as shown in Figure 311. This distance T is the amount by which car A must be translated downward in order to successfully turn out of the parking space.

The method to leave the parking space, shown in the top left corner of Figure 311, requires two phases to be successful: the initial **turning phase**, and the **straightening phase**. By turning and straightening out, we achieve a vertical shift downward and a horizontal shift left, while preserving the original orientation. That last part is key because if we attempted to turn until the corner of car A touched car B, car A would be rotated, and any attempt to straighten out would just follow the same arc backward to the initial position, while turning the wheel the other direction would rotate the car even more, as in the first method described above.

Our goal is to turn as far as we can and still be able to completely straighten out by time car A touches car B. To analyse just how much this turn should be, we must first look at the properties of a turning car.

Ackermann steering is the principle that in order for a car to turn smoothly, all four wheels must rotate about the same point. This was patented by Rudolph Ackermann in 1817. Some properties of Ackermann steering in relation to this problem are as follows:

- The back wheels stay in alignment, but the front wheels (which we control), must turn different amounts to rotate about the same centre.

- The centres of rotation for left and right turns are on opposite sides of the car

- For equal magnitudes of left and right turns, the centres of rotation are equidistant from the nearest edge of the car. Figure 311 makes this much clearer.

- All possible centres of rotation are on the same line, which also always passes through the back wheels.

- When the back wheels are 'straight' (straight will always mean in the same orientation as the initial position), they will be vertically aligned with the centres of rotation.

- When the car is turning about one centre, say the one associated with the maximum left turn, then the potential centre associated with the maximum right turn will rotate along with the car. Similarly, when the cars turns about the right centre, the left centre rotates.

Now that we know the properties of Ackermann steering, we can say that in order to maximize the shift downward while preserving the orientation, we must turn left about the 1st centre such that the 2nd centre rotates a *horizontal* distance d, as shown in Figure 311. When this is achieved, we brake, and turn the steering wheel the complete opposite direction so that we are now turning right about the 2nd centre. Because we shifted leftward d, we will straighten out at the exact moment car A comes in contact with car B. This results in our goal, a downward shift m and leftward shift d while preserving the orientation of car A. A similar process can be performed in reverse to achieve another downward shift m and a *rightward* shift d, effectively moving car A from its initial position (before any movement) downward $2m$ while preserving its orientation. This can be done indefinitely, which is why it is possible to get out of a parking space with an infinitesimal d between car A and car B. To determine how many times this procedure (both sets of turning and straightening) must be performed, we must only divide T (remember T is the amount by which car A must be shifted downward in order to turn out of the parking spot normally) by $2m$, the total downward shift for one iteration of the procedure. Symbolically,

$$n = \frac{T}{2m} .$$ (132)

In order to get an expression for n in terms of the geometry of the car, we must solve for T and

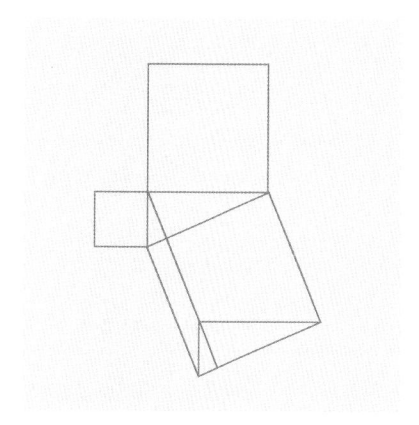

FIGURE 312 A simple drawing – one of the many possible one – that allows proving Pythagoras' theorem.

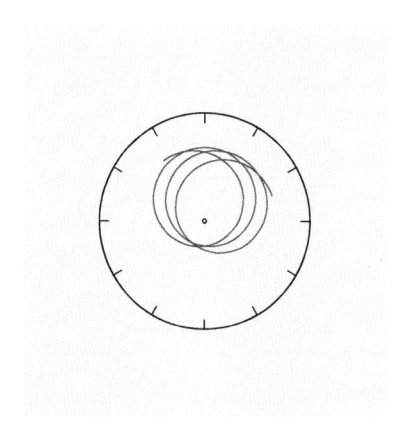

FIGURE 313 The trajectory of the middle point between the two ends of the hands of a clock.

$2m$. To simplify the derivations we define a new length x, also shown in Figure 311.

$$x = \sqrt{R^2 - (L - b)^2}$$

$$T = \sqrt{R^2 - (L - b + d)^2} - x + w$$
$$= \sqrt{R^2 - (L - b + d)^2} - \sqrt{R^2 - (L - b)^2} + w$$

$$m = 2x - w - \sqrt{(2x - w)^2 - d^2}$$
$$= 2\sqrt{R^2 - (L - b)^2} - w - \sqrt{(2\sqrt{R^2 - (L - b)^2} - w)^2 - d^2}$$
$$= 2\sqrt{R^2 - (L - b)^2} - w - \sqrt{4(R^2 - (L - b)^2) - 4w\sqrt{R^2 - (L - b)^2} + w^2 - d^2}$$
$$= 2\sqrt{R^2 - (L - b)^2} - w - \sqrt{4R^2 - 4(L - b)^2 - 4w\sqrt{R^2 - (L - b)^2} + w^2 - d^2}$$

We then get

$$n = \frac{T}{2m} = \frac{\sqrt{R^2 - (L - b + d)^2} - \sqrt{R^2 - (L - b)^2} + w}{4\sqrt{R^2 - (L - b)^2} - 2w - 2\sqrt{4R^2 - 4(L - b)^2 - 4w\sqrt{R^2 - (L - b)^2} + w^2 - d^2}} \, .$$

The value of n must always be rounded *up* to the next integer to determine how many times one must go backward and forward to leave the parking spot.

Challenge 102, page 66: Nothing, neither a proof nor a disproof.

Challenge 103, page 66: See volume II, on page 20. On extreme shutters, see also the discussion in Volume VI, on page 119.

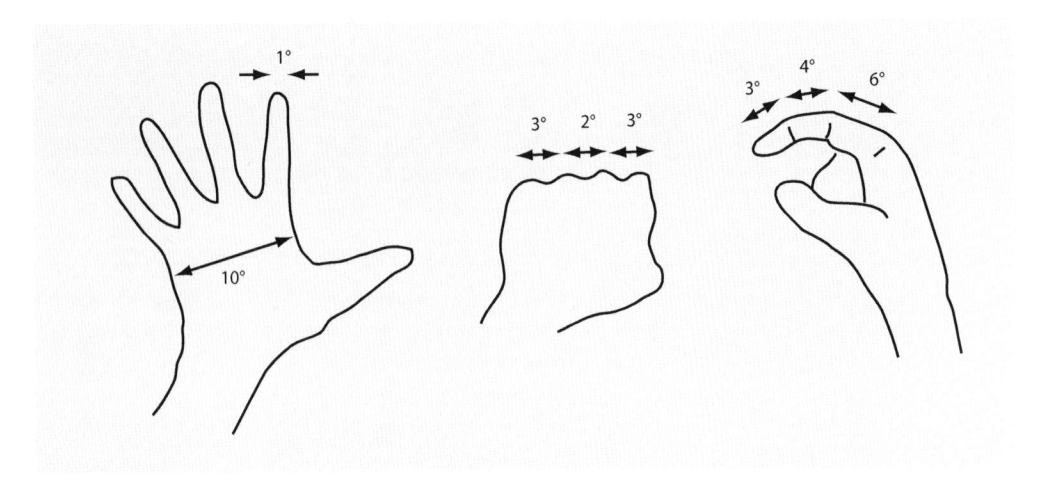

FIGURE 314 The angles defined by the hands against the sky, when the arms are extended.

Challenge 104, page 67: A hint for the solution is given in Figure 312.

Challenge 105, page 67: Because they are or were liquid.

Challenge 106, page 67: The shape is shown in Figure 313; it has eleven lobes.

Challenge 107, page 67: The cone angle φ, the angle between the cone axis and the cone border (or equivalently, *half* the apex angle of the cone) is related to the solid angle Ω through the relation $\Omega = 2\pi(1 - \cos\varphi)$. Use the surface area of a spherical cap to confirm this result.

Challenge 109, page 67: See Figure 314.

Challenge 113, page 69: Hint: draw all objects involved.

Challenge 114, page 69: The curve is obviously called a *catenary*, from Latin 'catena' for chain. The formula for a catenary is $y = a\cosh(x/a)$. If you approximate the chain by short straight segments, you can make wooden blocks that can form an arch without any need for glue. The St. Louis arch is in shape of a catenary. A suspension bridge has the shape of a catenary before it is loaded, i.e., before the track is attached to it. When the bridge is finished, the shape is in between a catenary and a parabola.

Challenge 115, page 69: The inverse radii, or curvatures, obey $a^2+b^2+c^2+d^2 = (1/2)(a+b+c+d)^2$. This formula was discovered by René Descartes. If one continues putting circles in the remaining spaces, one gets so-called circle packings, a pretty domain of recreational mathematics. They have many strange properties, such as intriguing relations between the coordinates of the circle centres and their curvatures.

Challenge 116, page 70: One option: use the three-dimensional analogue of Pythagoras's theorem. The answer is 9.

Challenge 117, page 70: There are two solutions. (Why?) They are the two positive solutions of $l^2 = (b+x)^2 + (b+b^2/x)^2$; the height is then given as $h = b + x$. The two solutions are 4.84 m and 1.26 m. There are closed formulas for the solutions; can you find them?

Challenge 118, page 70: The best way is to calculate first the height B at which the blue ladder touches the wall. It is given as a solution of $B^4 - 2hB^3 - (r^2 - b^2)B^2 + 2h(r^2 - b^2)B - h^2(r^2 - b^2) = 0$. Integer-valued solutions are discussed in MARTIN GARDNER, *Mathematical Circus*, Spectrum, 1996.

Challenge 119, page 70: Draw a logarithmic scale, i.e., put every number at a distance corresponding to its natural logarithm. Such a device, called a *slide rule*, is shown in Figure 315. Slide

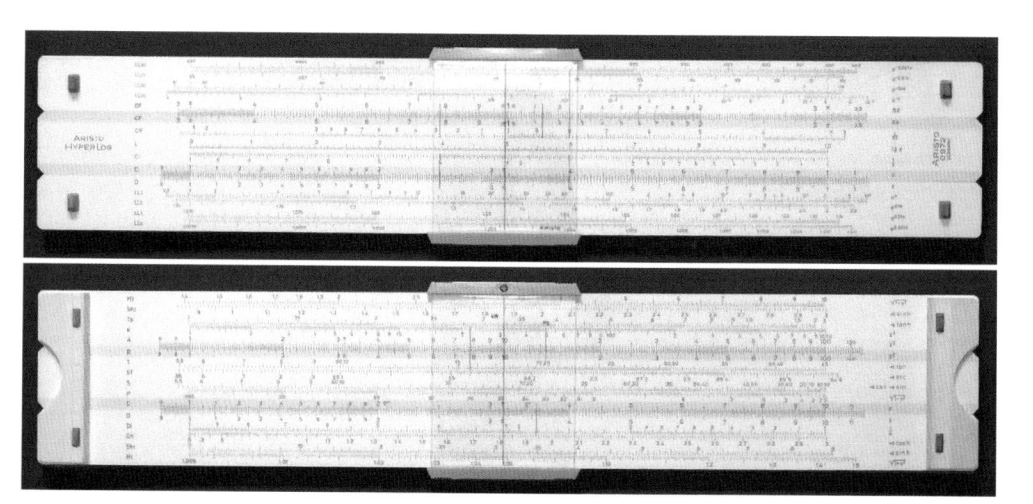

FIGURE 315 A high-end slide rule, around 1970 (© Jörn Lütjens).

rules were the precursors of electronic calculators; they were used all over the world in prehistoric times, i.e., until around 1970. See also the web page www.oughtred.org.

Challenge 120, page 71: Two more days. Build yourself a model of the Sun and the Earth to verify this. In fact, there is a small correction to the value 2, for the same reason that the makes the solar day shorter than 24 hours.

Challenge 121, page 71: The Sun is exactly behind the back of the observer; it is setting, and the rays are coming from behind and reach deep into the sky in the direction opposite to that of the Sun.

Challenge 123, page 71: The volume is given by $V = \int A dx = \int_{-1}^{1} 4(1 - x^2) dx = 16/3$.

Challenge 124, page 71: Yes. Try it with a paper model.

Challenge 125, page 71: Problems appear when quantum effects are added. A two-dimensional universe would have no matter, since matter is made of spin 1/2 particles. But spin 1/2 particles do not exist in two dimensions. Can you find additional reasons?

Challenge 126, page 72: Two dimensions of time do not allow ordering of events and observations. To say 'before' and 'afterwards' becomes impossible. In everyday life and all domains accessible to measurement, time is definitely one-dimensional.

Challenge 127, page 72: No experiment has ever found any hint. Can this be nevertheless? Probably not, as argued in the last volume of this book series.

Challenge 130, page 73: The best solution seems to be 23 extra lines. Can you deduce it? To avoid spoiling the fun of searching, no solution is given here. You can find solutions on blog.vixra.org/2010/12/26/a-christmas-puzzle.

Challenge 131, page 73: If you solve this so-called ropelength problem, you will become a famous mathematician. The length is known only with about 6 decimals of precision. No exact formula is known, and the exact shape of such ideal knots is unknown for all non-trivial knots. The problem is also unsolved for all non-trivial ideal *closed* knots, for which the two ends are glued together.

Challenge 132, page 76: From $x = gt^2/2$ you get the following rule: square the number of seconds, multiply by five and you get the depth in metres.

Challenge 133, page 76: Just experiment.

Challenge 134, page 77: The Academicians suspended one cannon ball with a thin wire just in front of the mouth of the cannon. When the shot was released, the second, flying cannon ball flew through the wire, thus ensuring that both balls started at the same time. An observer from far away then tried to determine whether both balls touched the Earth at the same time. The experiment is not easy, as small errors in the angle and air resistance confuse the results.

Challenge 135, page 77: A parabola has a so-called focus or focal point. All light emitted from that point and reflected exits in the same direction: all light rays are emitted in parallel. The name 'focus' – Latin for fireplace – expresses that it is the hottest spot when a parabolic mirror is illuminated. Where is the focus of the parabola $y = x^2$? (Ellipses have two foci, with a slightly different definition. Can you find it?)

Challenge 136, page 78: The long jump record could surely be increased by getting rid of the sand stripe and by measuring the true jumping distance with a photographic camera; that would allow jumpers to run more closely to their top speed. The record could also be increased by a small inclined step or by a spring-suspended board at the take-off location, to increase the take-off angle.

Challenge 137, page 78: It may be held by Roald Bradstock, who threw a golf ball over 155 m. Records for throwing mobile phones, javelins, people and washing machines are shorter.

Challenge 138, page 79: Walk or run in the rain, measure your own speed v and the angle from the vertical α with which the rain appears to fall. Then the speed of the rain is $v_{rain} = v/\tan \alpha$.

Challenge 139, page 79: In ice skating, quadruple jumps are now state of the art. In dance, no such drive for records exists.

Challenge 140, page 79: Neglecting air resistance and approximating the angle by 45°, we get $v = \sqrt{dg}$, or about 3.8 m/s. This speed is created by a steady pressure build-up, using blood pressure, which is suddenly released with a mechanical system at the end of the digestive canal. The cited reference tells more about the details.

Challenge 141, page 79: On horizontal ground, for a speed v and an angle from the horizontal α, neglecting air resistance and the height of the thrower, the distance d is $d = v^2 \sin 2\alpha/g$.

Challenge 142, page 79: Astonishingly, the answer is not clear. In 2012, the human record is eleven balls. For robots, the present record is three balls, as performed by the Sarcoman robot. The internet is full of material and videos on the topic. It is a challenge for people and robots to reach the maximum possible number of balls.

Challenge 143, page 79: It is said so, as rain drops would then be ice spheres and fall with high speed.

Challenge 144, page 80: Yes! People have gone to hospital and even died because a falling bullet went straight through their head. See S. MIRSKY, *It is high, it is far*, Scientific American p. 86, February 2004, or C. TUIJN, *Vallende kogels*, Nederlands tijdschrift voor natuurkunde 71, pp. 224–225, 2005. Firing a weapon into the air is a crime.

Challenge 145, page 80: This is a true story. The answer can only be given if it is known whether the person had the chance to jump while running or not. In the case described by R. CROSS, *Forensic physics 101: falls from a height*, American Journal of Physics 76, pp. 833–837, 2008, there was no way to run, so that the answer was: murder.

Challenge 146, page 80: For jumps of an animal of mass m the necessary energy E is given as $E = mgh$, and the work available to a muscle is roughly speaking proportional to its mass $W \sim m$. Thus one gets that the height h is independent of the mass of the animal. In other words, the specific mechanical energy of animals is around 1.5 ± 0.7 J/kg.

Challenge 147, page 80: Stones *never* follow parabolas: when studied in detail, i.e., when the change of g with height is taken into account, their precise path turns out to be an ellipse. This shape appears most clearly for long throws, such as throws around a sizeable part of the Earth, or for orbiting objects. In short, stones follow parabolas only if the Earth is assumed to be flat. If its curvature is taken into account, they follow ellipses.

Challenge 148, page 81: The set of all rotations around a point in a plane is indeed a vector space. What about the set of all rotations around *all* points in a plane? And what about the three-dimensional cases?

Challenge 151, page 81: The scalar product between two vectors *a* and *b* is given by

$$ab = ab \cos \sphericalangle(\boldsymbol{a}, \boldsymbol{b}) \; . \tag{133}$$

How does this differ from the vector product?

Challenge 154, page 84: One candidate for the lowest practical acceleration of a physical system are the accelerations measured by gravitational wave detectors. They are below 10^{-13} m/s^2. But these low values are beaten by the acceleration of the continental drift after the continents 'snap' apart: they accelerate from 7 mm/a to 40 mm/a in a 'mere' 3 million years. This corresponds to a value of 10^{-23} m/s^2. Is there a theoretical lowest limit to acceleration?

Challenge 155, page 85: In free fall (when no air is present) or inside a space station orbiting the Earth, you are accelerated but do not feel anything. However, the issue is not so simple. On the one hand, constant and homogeneous accelerations are indeed not felt if there is no non-accelerated reference. This indistinguishability or equivalence between acceleration and 'feeling nothing' was an essential step for Albert Einstein in his development of general relativity. On the other hand, if our senses were sensitive enough, we would feel something: both in the free fall and in the space station, the acceleration is neither constant nor homogeneous. So we can indeed say that accelerations found in nature can always be felt.

Challenge 156, page 85: Professor to student: What is the derivative of velocity? Acceleration! What is the derivative of acceleration? I don't know. *Jerk!* The fourth, fifth and sixth derivatives of position are sometimes called *snap*, *crackle* and *pop*.

Challenge 158, page 87: One can argue that any source of light must have finite size.

Challenge 160, page 89: What the unaided human eye perceives as a tiny black point is usually about 50 μm in diameter.

Challenge 161, page 89: See volume III, page 170.

Challenge 162, page 89: One has to check carefully whether the conceptual steps that lead us to extract the concept of point from observations are correct. It will be shown in the final part of the adventure that this is not the case.

Challenge 163, page 89: One can rotate the hand in a way that the arm makes the motion described. See also volume IV, page 132.

Challenge 164, page 89: The number of cables has no limit. A visualization of tethered rotation with 96 connections is found in volume VI, on page 181.

Challenge 165, page 90: The blood and nerve supply is not possible if the wheel has an axle. The method shown to avoid tangling up connections only works when the rotating part has *no* axle: the 'wheel' must float or be kept in place by other means. It thus becomes impossible to make a wheel *axle* using a single piece of skin. And if a wheel without an axle could be built (which might be possible), then the wheel would periodically run over the connection. Could such a axle-free connection realize a propeller?

By the way, it is still thinkable that animals have wheels on axles, if the wheel is a 'dead' object. Even if blood supply technologies like continuous flow reactors were used, animals could not make such a detached wheel grow in a way tuned to the rest of the body and they would have difficulties repairing a damaged wheel. Detached wheels cannot be grown on animals; they must be dead.

Challenge 166, page 92: The brain in the skull, the blood factories inside bones or the growth of the eye are examples.

Challenge 167, page 92: In 2007, the largest big wheels for passengers are around 150 m in diameter. The largest wind turbines are around 125 m in diameter. Cement kilns are the longest wheels: they can be over 300 m along their axis.

Challenge 168, page 92: Air resistance reduces the maximum distance achievable with the soccer ball – which is realized for an angle of about $\pi/4 = 45°$ – from around $v^2/g = 91.7$ m down to around 50 m.

Challenge 173, page 98: One can also add the Sun, the sky and the landscape to the list.

Challenge 174, page 99: There is no third option. Ghosts, hallucinations, Elvis sightings, or extraterrestrials must all be objects or images. Also shadows are only special types of images.

Challenge 175, page 99: The issue was hotly discussed in the seventeenth century; even Galileo argued for them being images. However, they are objects, as they can collide with other objects, as the spectacular collision between Jupiter and the comet Shoemaker-Levy 9 in 1994 showed. In the meantime, satellites have been made to collide with comets and even to shoot at them (and hitting).

Challenge 176, page 100: The minimum speed is roughly the one at which it is possible to ride without hands. If you do so, and then *gently* push on the steering wheel, you can make the experience described above. Watch out: too strong a push will make you fall badly.

The *bicycle* is one of the most complex mechanical systems of everyday life, and it is still a subject of research. And obviously, the world experts are Dutch. An overview of the behaviour of a bicycle is given in Figure 316. The main result is that the bicycle is stable in the upright position at a range of medium speeds. Only at low and at large speeds must the rider actively steer to ensure upright position of the bicycle.

For more details, see the paper J. P. Meijaard, J. M. Papadopoulos, A. Ruina & A. L. Schwab, *Linearized dynamics equations for the balance and steer of a bicycle: a benchmark and review*, Proceedings of the Royal Society A 463, pp. 1955–1982, 2007, and J. D. G. Kooijman, A. L. Schwab & J. P. Meijaard, *Experimental validation of a model of an uncontrolled bicycle*, Multibody System Dynamics 19, pp. 115–132, 2008. See also the audiophile.tam.cornell.edu/~als93/Bicycle/index.htm website.

Challenge 177, page 102: The total weight decreased slowly, due to the evaporated water lost by sweating and, to a minor degree, due to the exhaled carbon bound in carbon dioxide.

Challenge 178, page 102: This is a challenge where the internet can help a lot. For a general introduction, see the book by Lee Siegel, *Net of Magic – Wonders and Deception in India*, University of Chicago Press, 1991.

Challenge 179, page 103: If the moving ball is not rotating, after the collision the two balls will depart with a *right* angle between them.

Challenge 180, page 103: As the block is heavy, the speed that it acquires from the hammer is small and easily stopped by the human body. This effect works also with an anvil instead of a concrete block. In another common variation the person does not lie on nails, but on air: he just keeps himself horizontal, with head and shoulders on one chair, and the feet on a second one.

Challenge 181, page 104: Yes, the definition of mass works also for magnetism, because the precise condition is not that the interaction is central, but that the interaction realizes a more general condition that includes accelerations such as those produced by magnetism. Can you deduce the condition from the definition of mass as that quantity that keeps momentum conserved?

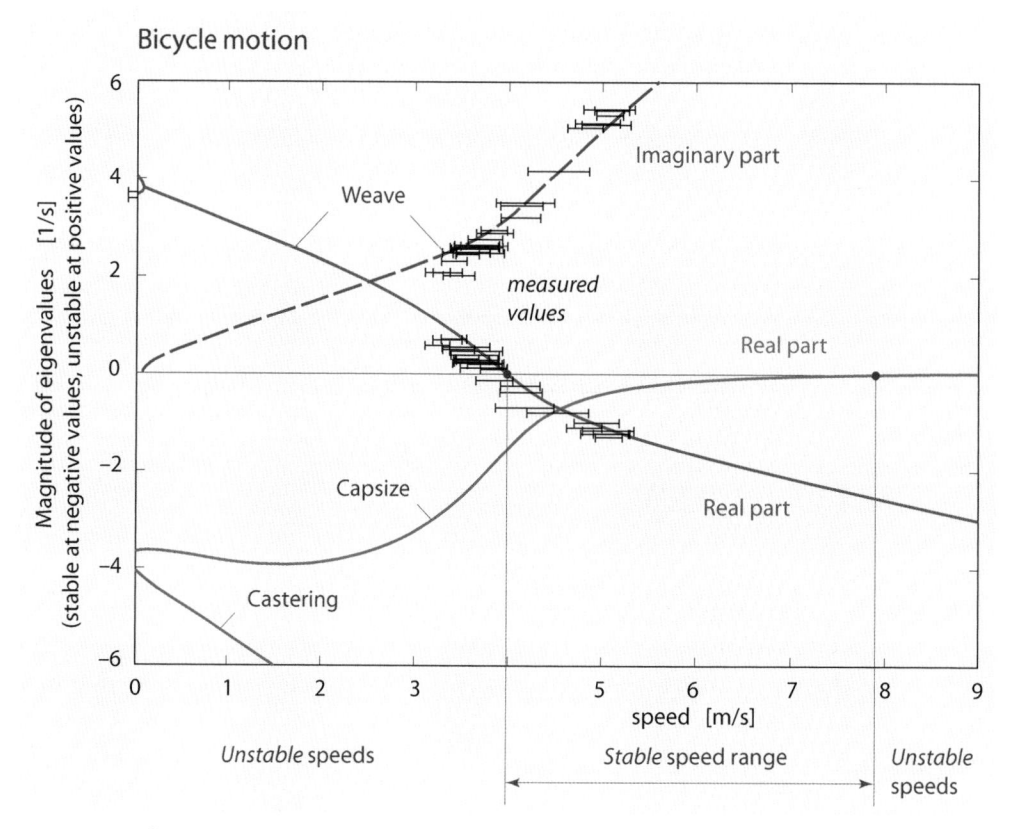

FIGURE 316 The measured (black bars) and calculated behaviour (coloured lines) – more precisely, the dynamical eigenvalues – of a bicycle as a function of its speed (© Arend Schwab).

Challenge 182, page 106: Rather than using the inertial effects of the Earth, it is easier to deduce its mass from its gravitational effects.

Page 185

Challenge 187, page 107: At first sight, relativity implies that tachyons have imaginary mass; however, the imaginary factor can be extracted from the mass–energy and mass–momentum relation, so that one can define a real mass value for tachyons. As a result, faster tachyons have smaller energy and smaller momentum than slower ones. In fact, both tachyon momentum and tachyon energy can be a negative number of any size.

Challenge 188, page 109: The leftmost situation has a tiny effect, the second situation makes the car roll forward and backward, the right two pictures show ways to open wine bottles without bottle opener.

Challenge 189, page 109: Legs are never perfectly vertical; they would immediately glide away. Once the cat or the person is on the floor, it is almost impossible to stand up again.

Challenge 190, page 109: Momentum (or centre of mass) conservation would imply that the environment would be accelerated into the opposite direction. Energy conservation would imply that a huge amount of energy would be transferred between the two locations, melting everything in between. Teleportation would thus contradict energy and momentum conservation.

Challenge 191, page 110: The part of the tides due to the Sun, the solar wind, and the interactions between both magnetic fields are examples of friction mechanisms between the Earth and the Sun.

Challenge 192, page 110: With the factor 1/2, increase of (physical) kinetic energy is equal to the (physical) work performed on a system: total energy is thus conserved only if the factor 1/2 is added.

Challenge 194, page 112: It is a smart application of momentum conservation.

Challenge 195, page 112: Neither. With brakes on, the damage is higher, but still equal for both cars.

Challenge 196, page 113: Heating systems, transport engines, engines in factories, steel plants, electricity generators covering the losses in the power grid, etc. By the way, the richest countries in the world, such as Sweden or Switzerland, consume only half the energy per inhabitant as the USA. This waste is one of the reasons for the lower average standard of living in the USA.

Challenge 202, page 117: Just throw the brick into the air and compare the dexterity needed to make it turn around various axes.

Challenge 203, page 118: Use the definition of the moment of inertia and Pythagoras' theorem for every mass element of the body.

Challenge 204, page 119: Hang up the body, attaching the rope in two different points. The crossing point of the prolonged rope lines is the centre of mass.

Challenge 205, page 119: See Tables 19 and 20.

Challenge 206, page 119: Spheres have an orientation, because we can always add a tiny spot on their surface. This possibility is not given for microscopic objects, and we shall study this situation in the part on quantum theory.

Challenge 209, page 120: Yes, the ape can reach the banana. The ape just has to turn around its own axis. For every turn, the plate will rotate a bit towards the banana. Of course, other methods, like blowing at a right angle to the axis, peeing, etc., are also possible.

Challenge 210, page 121: Self-propelled linear motion contradicts the conservation of momentum; self-propelled change of orientation (as long as the motion stops again) does not contradict any conservation law. But the deep, underlying reason for the difference will be unveiled in the final part of our adventure.

Challenge 212, page 121: The points that move exactly along the radial direction of the wheel form a circle below the axis and above the rim. They are the points that are sharp in Figure 79 of page 121.

Challenge 213, page 122: Use the conservation of angular momentum around the point of contact. If all the wheel's mass is assumed in the rim, the final rotation speed is half the initial one; it is independent of the friction coefficient.

Challenge 215, page 123: Probably the 'rest of the universe' was meant by the writer. Indeed, a moving a part never shifts the centre of gravity of a closed system. But is the universe closed? Or a system? The last part of our adventure addresses these issues.

Challenge 219, page 124: Hint: energy and momentum conservation yield two equations; but in the case of three balls there are three variables. What else is needed? See F. HERRMANN & M. SEITZ, *How does the ball-chain work?*, American Journal of Physics 50, pp. 977–981, 1982. The bigger challenge is to build a high precision ball-chain, in which the balls behave as expected, minimizing spurious motion. Nobody seems to have built one yet, as the internet shows. Can you?

Challenge 220, page 124: The method allowed Phileas Fogg to win the central bet in the well-known adventure novel by JULES VERNE, *Around the World in Eighty Days*, translated from *Le tour du monde en quatre-vingts jours*, first published in 1872.

Challenge 221, page 125: The human body is more energy-efficient at low and medium power output. The topic is still subject of research, as detailed in the cited reference. The critical slope is estimated to be around 16° for uphill walkers, but should differ for downhill walkers.

Challenge 223, page 125: Hint: an energy per distance is a force.

Challenge 224, page 126: The conservation of angular momentum saves the glass. Try it.

Challenge 225, page 126: First of all, MacDougall's experimental data is flawed. In the six cases MacDougall examined, he did not know the exact timing of death. His claim of a mass decrease cannot be deduced from his own data. Modern measurements on dying sheep, about the same mass as humans, have shown no mass change, but clear weight pulses of a few dozen grams when the heart stopped. This temporary weight decrease could be due to the expelling of air or moisture, to the relaxing of muscles, or to the halting of blood circulation. The question is not settled.

Challenge 227, page 126: Assuming a square mountain, the height h above the surrounding crust and the depth d below are related by

$$\frac{h}{d} = \frac{\rho_m - \rho_c}{\rho_c} \tag{134}$$

where ρ_c is the density of the crust and ρ_m is the density of the mantle. For the density values given, the ratio is 6.7, leading to an additional depth of 6.7 km below the mountain.

Challenge 229, page 127: The can filled with liquid. Videos on the internet show the experiment. Why is this the case?

Challenge 232, page 127: The matter in the universe could rotate – but not the universe itself. Measurements show that within measurement errors there is no mass rotation.

Challenge 233, page 128: The behaviour of the spheres can only be explained by noting that elastic waves propagate through the chain of balls. Only the propagation of these elastic waves, in particular their reflection at the end of the chain, explains that the same number of balls that hit on one side are lifted up on the other. For long times, friction makes all spheres oscillate in phase. Can you confirm this?

Challenge 234, page 128: When the short cylinder hits the long one, two compression waves start to run from the point of contact through the two cylinders. When each compression wave arrives at the end, it is reflected as an expansion wave. If the geometry is well chosen, the expansion wave coming back from the short cylinder can continue into the long one (which is still in his compression phase). For sufficiently long contact times, waves from the short cylinder can thus depose much of their energy into the long cylinder. Momentum is conserved, as is energy; the long cylinder is oscillating in length when it detaches, so that not all its energy is translational energy. This oscillation is then used to drive nails or drills into stone walls. In commercial hammer drills, length ratios of 1:10 are typically used.

Challenge 235, page 128: The momentum transfer to the wall is double when the ball rebounds perfectly.

Challenge 236, page 128: If the cork is in its intended position: take the plastic cover off the cork, put the cloth around the bottle or the bottle in the shoe (this is for protection reasons only) and repeatedly hit the bottle on the floor or a fall in an inclined way, as shown in Figure 72 on page 109. With each hit, the cork will come out a bit.

If the cork has fallen inside the bottle: put half the cloth inside the bottle; shake until the cork falls unto the cloth. Pull the cloth out: first slowly, until the cloth almost surround the cork, and then strongly.

Challenge 237, page 129: Indeed, the lower end of the ladder always touches the floor. Why?

Challenge 238, page 129: The atomic force microscope.

Challenge 240, page 129: Running man: $E \approx 0.5 \cdot 80\,\text{kg} \cdot (5\,\text{m/s})^2 = 1\,\text{kJ}$; rifle bullet: $E \approx 0.5 \cdot 0.04\,\text{kg} \cdot (500\,\text{m/s})^2 = 5\,\text{kJ}$.

Challenge 241, page 129: It almost doubles in size.

Challenge 242, page 130: At the highest point, the acceleration is $g \sin \alpha$, where α is the angle of the pendulum at the highest point. At the lowest point, the acceleration is v^2/l, where l is the length of the pendulum. Conservation of energy implies that $v^2 = 2gl(1 - \cos \alpha)$. Thus the problem requires that $\sin \alpha = 2(1 - \cos \alpha)$. This results in $\cos \alpha = 3/5$.

Challenge 243, page 130: One needs the mass change equation $dm/dt = \pi \rho_{\text{vapour}} r^2 |v|$ due to the mist and the drop speed evolution $m\,dv/dt = mg - v\,dm/dt$. These two equations yield

$$\frac{dv^2}{dr} = \frac{2g}{C} - 6\frac{v^2}{r} \tag{135}$$

where $C = \rho_{\text{vapour}}/4\rho_{\text{water}}$. The trick is to show that this can be rewritten as

$$r\frac{d}{dr}\frac{v^2}{r} = \frac{2g}{C} - 7\frac{v^2}{r}\,. \tag{136}$$

For large times, all physically sensible solutions approach $v^2/r = 2g/7C$; this implies that for large times,

$$\frac{dv}{dt}\frac{v^2}{r} = \frac{g}{7} \quad \text{and} \quad r = \frac{gC}{14}t^2\,. \tag{137}$$

About this famous problem, see for example, B. F. EDWARDS, J. W. WILDER & E. E. SCIME, *Dynamics of falling raindrops*, European Journal of Physics **22**, pp. 113–118, 2001, or A. D. SOKAL, *The falling raindrop, revisited*, preprint at arxiv.org/abs/0908.0090.

Challenge 244, page 130: One is faster, because the moments of inertia differ. Which one?

Challenge 245, page 130: There is no simple answer, as aerodynamic drag plays an important role. There are almost no studies on the topic. By the way, competitive rope jumping is challenging; for example, a few people in the world are able to rotate the rope 5 times under their feet during a single jump. Can you do better?

Challenge 246, page 130: Weigh the bullet and shoot it against a mass hanging from the ceiling. From the mass and the angle it is deflected to, the momentum of the bullet can be determined.

Challenge 248, page 130: The curve described by the midpoint of a ladder sliding down a wall is a circle.

Challenge 249, page 130: The switches use the power that is received when the switch is pushed and feed it to a small transmitter that acts a high frequency remote control to switch on the light.

Challenge 250, page 131: A clever arrangement of bimetals is used. They move every time the temperature changes from day to night – and vice versa – and wind up a clock spring. The clock itself is a mechanical clock with low energy consumption.

Challenge 251, page 131: The weight of the lift does not change at all when a ship enters it. A twin lift, i.e., a system in which both lifts are mechanically or hydraulically connected, needs no engine at all: it is sufficient to fill the upper lift with a bit of additional water every time a ship enters it. Such ship lifts without engines at all used to exist in the past.

Challenge 254, page 133: This is not easy; a combination of friction and torques play a role. See for example the article J. SAUER, E. SCHÖRNER & C. LENNERZ, *Real-time rigid body simulation of some classical mechanical toys*, 10th European Simulation and Symposium and Exhibition (ESS '98) 1998, pp. 93–98, or www.lennerz.de/paper_ess98.pdf.

Page 166

Challenge 257, page 135: See Figure 123 for an example. The pole is not at the zenith.

Challenge 258, page 135: Robert Peary had forgotten that on the date he claimed to be at the North Pole, 6th of April 1909, the Sun is very low on the horizon, casting very long shadows, about ten times the height of objects. But on his photograph the shadows are much shorter. (In fact, the picture is taken in such a way to hide all shadows as carefully as possible.) Interestingly, he had even convinced the US congress to officially declare him the first man on the North Pole in 1911. (A rival crook had claimed to have reached it before Peary, but his photograph has the same mistake.) Peary also cheated on the travelled distances of the last few days; he also failed to mention that the last days he was pulled by his partner, Matthew Henson, because he was not able to walk any more. In fact Matthew Henson deserves more credit for that adventure than Peary. Henson, however, did not know that Peary cheated on the position they had reached.

Challenge 260, page 136: Laplace and Gauss showed that the eastward deflection d of a falling object is given by

$$d = 2/3\Omega \cos \varphi \sqrt{2h^3/g} \ . \tag{138}$$

Here $\Omega = 72.92 \,\mu\text{rad/s}$ is the angular velocity of the Earth, φ is the latitude, g the gravitational acceleration and h is the height of the fall.

Challenge 261, page 139: The Coriolis effect can be seen as the sum two different effects of equal magnitude. The first effect is the following: on a rotating background, velocity changes over time. What an inertial (non-rotating) observer sees as a *constant* velocity will be seen a velocity that *changes* over time by the rotating observer. The acceleration seen by the rotating observer is negative, and is proportional to the angular velocity and to the velocity.

The second effect is change of velocity in space. In a rotating frame of reference, different points have different velocities. The effect is negative, and proportional to the angular velocity and to the velocity.

In total, the Coriolis acceleration (or Coriolis effect) is thus $\boldsymbol{a}_C = -2\boldsymbol{\omega} \times \boldsymbol{v}$.

Challenge 262, page 140: A *short* pendulum of length L that swings in two dimensions (with amplitude ρ and orientation φ) shows two additional terms in the Lagrangian $\mathcal{L}$:

$$\mathcal{L} = T - V = \tfrac{1}{2}m\dot{\rho}^2\left(1 + \frac{\rho^2}{L^2}\right) + \frac{l_z^2}{2m\rho^2} - \tfrac{1}{2}m\omega_0^2\rho^2(1 + \frac{\rho^2}{4\,L^2}) \tag{139}$$

where as usual the basic frequency is $\omega_0^2 = g/L$ and the angular momentum is $l_z = m\rho^2\dot{\varphi}$. The two additional terms disappear when $L \to \infty$; in that case, if the system oscillates in an ellipse with semiaxes a and b, the ellipse is fixed in space, and the frequency is ω_0. For *finite* pendulum length L, the frequency changes to

$$\omega = \omega_0\left(1 - \frac{a^2 + b^2}{16\,L^2}\right) \ . \tag{140}$$

The ellipse turns with a frequency

$$\Omega = \omega\frac{3}{8}\frac{ab}{L^2} \ . \tag{141}$$

These formulae can be derived using the least action principle, as shown by C. G. GRAY, G. KARL & V. A. NOVIKOV, *Progress in classical and quantum variational principles*, arxiv.org/abs/physics/0312071. In other words, a short pendulum in elliptical motion shows a precession even *without* the Coriolis effect. Since this precession frequency diminishes with $1/L^2$, the effect is small for long pendulums, where only the Coriolis effect is left over. To see the Coriolis effect in a short pendulum, one thus has to avoid that it starts swinging in an elliptical orbit by adding a mechanism that suppresses elliptical motion.

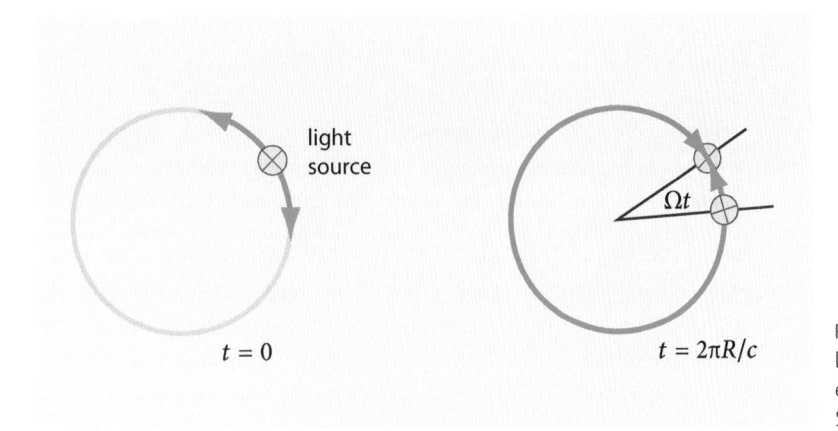

FIGURE 317
Deducing the
expression for the
Sagnac effect.

Challenge 263, page 141: The Coriolis acceleration is the reason for the deviation from the straight line. The Coriolis acceleration is due to the change of speed with distance from the rotation axis. Now think about a pendulum, located in Paris, swinging in the North-South direction with amplitude A. At the Southern end of the swing, the pendulum is further from the axis by $A \sin \varphi$, where φ is the latitude. At that end of the swing, the central support point overtakes the pendulum bob with a relative horizontal speed given by $v = 2\pi A \sin \varphi / 23\,\mathrm{h}56\,\mathrm{min}$. The period of precession is given by $T_\mathrm{F} = v/2\pi A$, where $2\pi A$ is the circumference $2\pi A$ of the envelope of the pendulum's path (relative to the Earth). This yields $T_\mathrm{F} = 23\,\mathrm{h}56\,\mathrm{min}/\sin\varphi$. Why is the value that appears in the formula not $24\,\mathrm{h}$, but $23\,\mathrm{h}56\,\mathrm{min}$?

Challenge 264, page 141: Experiments show that the axis of the gyroscope stays fixed with respect to distant stars. No experiment shows that it stays fixed with respect to absolute space, because this kind of "absolute space" cannot be defined or observed at all. It is a useless concept.

Challenge 265, page 141: Rotation leads to a small frequency and thus colour changes of the circulating light.

Challenge 266, page 141: The weight changes when going east or when moving west due to the Coriolis acceleration. If the rotation speed is tuned to the oscillation frequency of the balance, the effect is increased by resonance. This trick was also used by Eötvös.

Challenge 267, page 141: The Coriolis acceleration makes the bar turn, as every moving body is deflected to the side, and the two deflections add up in this case. The direction of the deflection depends on whether the experiments is performed on the northern or the southern hemisphere.

Challenge 268, page 141: When rotated by π around an east–west axis, the Coriolis force produces a drift velocity of the liquid around the tube. It has the value

$$v = 2\omega r \sin \theta, \tag{142}$$

as long as friction is negligible. Here ω is the angular velocity of the Earth, θ the latitude and r the (larger) radius of the torus. For a tube with $1\,\mathrm{m}$ diameter in continental Europe, this gives a speed of about $6.3 \cdot 10^{-5}\,\mathrm{m/s}$.

The measurement can be made easier if the tube is restricted in diameter at one spot, so that the velocity is increased there. A restriction by an area factor of 100 increases the speed by the same factor. When the experiment is performed, one has to carefully avoid any other effects that lead to moving water, such as temperature gradients across the system.

Challenge 269, page 142: Imagine a circular light path (for example, inside a circular glass fibre) and two beams moving in opposite directions along it, as shown in Figure 317. If the fibre path

rotates with rotation frequency Ω, we can deduce that, after one turn, the difference ΔL in path length is

$$\Delta L = 2R\Omega t = \frac{4\pi R^2 \Omega}{c} . \tag{143}$$

The phase difference is thus

$$\Delta\varphi = \frac{8\pi^2 R^2}{c\lambda}\Omega \tag{144}$$

if the refractive index is 1. This is the required formula for the main case of the Sagnac effect.

It is regularly suggested that the Sagnac effect can only be understood with help of general relativity; this is wrong. As just done, the effect is easily deduced from the invariance of the speed of light c. The effect is a consequence of special relativity.

Challenge 270, page 145: The metal rod is slightly longer on one side of the axis. When the wire keeping it up is burned with a candle, its moment of inertia decreases by a factor of 10^4; thus it starts to rotate with (ideally) 10^4 times the rotation rate of the Earth, a rate which is easily visible by shining a light beam on the mirror and observing how its reflection moves on the wall.

Challenge 272, page 152: The original result by Bessel was 0.3136 ″, or 657.7 thousand orbital radii, which he thought to be 10.3 light years or 97.5 Pm.

FIGURE 318 How the night sky, and our galaxy in particular, looks in the near infrared (NASA false colour image).

Challenge 274, page 155: The galaxy forms a stripe in the sky. The galaxy is thus a flattened structure. This is even clearer in the infrared, as shown more clearly in Figure 318. From the flattening (and its circular symmetry) we can deduce that the galaxy must be rotating. Thus other matter must exist in the universe.

Challenge 276, page 158: See page 189.

Challenge 278, page 158: The scale reacts to your heartbeat. The weight is almost constant over time, except when the heart beats: for a short duration of time, the weight is somewhat lowered at each beat. Apparently it is due to the blood hitting the aortic arch when the heart pumps it upwards. The speed of the blood is about 0.3 m/s at the maximum contraction of the left ventricle. The distance to the aortic arch is a few centimetres. The time between the contraction and the reversal of direction is about 15 ms. And the measured weight is not even constant for a dead person, as air currents disturb the measurement.

Challenge 279, page 158: Use Figure 97 on page 138 for the second half of the trajectory, and think carefully about the first half. The body falls down slightly to the west of the starting point.

Challenge 280, page 158: Hint: starting rockets at the Equator saves a lot of energy, thus of fuel and of weight.

Challenge 283, page 161: The flame leans towards the inside.

Challenge 284, page 161: Yes. There is no absolute position and no absolute direction. Equivalently, there is no preferred position, and no preferred direction. For time, only the big bang seems to provide an exception, at first; but when quantum effects are included, the lack of a preferred time scale is confirmed.

Challenge 285, page 161: For your exam it is better to say that centrifugal force does not exist. But since in each stationary system there is a force balance, the discussion is somewhat a red herring.

Challenge 288, page 162: Place the tea in cups on a board and attach the board to four long ropes that you keep in your hand.

Challenge 289, page 162: The ball leans in the direction it is accelerated to. As a result, one could imagine that the ball in a glass at rest pulls upwards because the floor is accelerated upwards. We will come back to this issue in the section of general relativity.

Challenge 290, page 162: The friction of the tides on Earth are the main cause.

Challenge 291, page 163: An earthquake with Richter magnitude of 12 is 1000 times the energy of the 1960 Chile quake with magnitude 10; the latter was due to a crack throughout the full 40 km of the Earth's crust along a length of 1000 km in which both sides slipped by 10 m with respect to each other. Only the impact of a large asteroid could lead to larger values than 12.

Challenge 293, page 163: Yes; it happens twice a year. To minimize the damage, dishes should be dark in colour.

Challenge 294, page 163: A rocket fired from the back would be a perfect defence against planes attacking from behind. However, when released, the rocket is effectively flying backwards with respect to the air, thus turns around and then becomes a danger to the plane that launched it. Engineers who did not think about this effect almost killed a pilot during the first such tests.

Challenge 295, page 163: Whatever the ape does, whether it climbs up or down or even lets himself fall, it remains at the same height as the mass. Now, what happens if there is friction at the wheel?

Challenge 296, page 163: Yes, if he moves at a large enough angle to the direction of the boat's motion.

Challenge 297, page 163: See the article by C. UCKE & H.-J. SCHLICHTING, *Faszinierendes Dynabee*, Physik in unserer Zeit **33**, pp. 230–231, 2002.

Challenge 298, page 164: See the article by C. UCKE & H.-J. SCHLICHTING, *Die kreisende Büroklammer*, Physik in unserer Zeit **36**, pp. 33–35, 2005.

Challenge 299, page 164: If a wedding ring rotates on an axis that is not a principal one, angular momentum and velocity are not parallel.

Challenge 300, page 164: The moment of inertia for a homogeneous sphere is $\Theta = \frac{2}{5}mr^2$.

Challenge 301, page 164: The three moments of inertia for the cube are equal, as in the case of the sphere, but the values are $\Theta = \frac{1}{6}ml^2$. The efforts required to put a sphere and a cube into rotationthe are thus different.

Challenge 304, page 165: Yes, the moon differs in this way. Can you imagine what happens for an observer on the Equator?

Challenge 305, page 167: A straight line at the zenith, and circles getting smaller at both sides. See an example on the website apod.nasa.gov/apod/ap021115.html.

Challenge 307, page 167: The plane is described in the websites cited; for a standing human the plane is the vertical plane containing the two eyes.

Challenge 308, page 168: If you managed, please send the author the video!

Challenge 309, page 169: As said before, legs are simpler than wheels to grow, to maintain and to repair; in addition, legs do not require flat surfaces (so-called 'streets') to work.

Challenge 310, page 170: The staircase formula is an empirical result found by experiment, used by engineers world-wide. Its origin and explanation seems to be lost in history.

Challenge 311, page 170: Classical or everyday nature is right-left symmetric and thus requires an even number of legs. Walking on two-dimensional surfaces naturally leads to a minimum of four legs. Starfish, snails, slugs, clams, eels and snakes are among the most important exceptions for which the arguments are not valid.

Challenge 313, page 173: The length of the day changes with latitude. So does the length of a shadow or the elevation of stars at night, facts that are easily checked by telephoning a friend. Ships appear at the horizon by showing their masts first. These arguments, together with the round shadow of the earth during a lunar eclipse and the observation that everything falls downwards everywhere, were all given already by Aristotle, in his text *On the Heavens*. It is now known that everybody in the last 2500 years knew that the Earth is a sphere. The myth that many people used to believe in a flat Earth was put into the world – as rhetorical polemic – by Copernicus. The story then continued to be exaggerated more and more during the following centuries, because a new device for spreading lies had just been invented: book printing. Fact is that for 2500 years Vol. III, page 313 the vast majority of people knew that the Earth is a sphere.

Challenge 314, page 177: The vector SF can be calculated by using $SC = -(GmM/E)\,SP/SP$ and then translating the construction given in the figure into formulae. This exercise yields

$$SF = \frac{K}{mE} \tag{145}$$

where

$$K = p \times L - GMm^2 x/x \tag{146}$$

is the so-called *Runge–Lenz vector*. The Runge-Lenz vector is directed along the line that connects the second focus to the first focus of the ellipse (the Sun). We have used $x = SP$ for the position of the orbiting body, p for its momentum and L for its angular momentum. The Runge–Lenz vector K is *constant* along the orbit of a body, thus has the *same* value for any position x on the orbit. (Prove it by starting from $xK = xK\cos\theta$.) The Runge–Lenz vector is thus a *conserved* quantity in universal gravity. As a result, the vector SF is also constant in time.

The Runge–Lenz vector is also often used in quantum mechanics, when calculating the energy levels of a hydrogen atom, as it appears in all problems with a $1/r$ potential. (In fact, the incorrect name 'Runge–Lenz vector' is due to Wolfgang Pauli; the discoverer of the vector was, in 1710, Jakob Hermann.)

Challenge 316, page 178: On orbits, see page 192.

Challenge 317, page 178: The low gravitational acceleration of the Moon, $1.6\,\text{m/s}^2$, implies that gas molecules at usual temperatures can escape its attraction.

Challenge 318, page 179: The tip of the velocity arrow, when drawn over time, produces a circle around the centre of motion.

Challenge 319, page 179: Draw a figure of the situation.

Challenge 320, page 179: Again, draw a figure of the situation.

Challenge 321, page 180: The value of the product GM for the Earth is $4.0 \cdot 10^{14}\,\text{m}^3/\text{s}^2$.

Challenge 322, page 180: All points can be reached for general inclinations; but when shooting horizontally in one given direction, only points on the first half of the circumference can be reached.

Challenge 323, page 182: On the moon, the gravitational acceleration is $1.6\,\text{m/s}^2$, about one sixth of the value on Earth. The surface values for the gravitational acceleration for the planets can be found on many internet sites.

Challenge 324, page 182: The Atwood machine is the answer: two almost equal masses m_1 and m_2 connected by a string hanging from a well-oiled wheel of negligible mass. The heavier one falls very slowly. Can show that the acceleration a of this 'unfree' fall is given by $a = g(m_1 - m_2)/(m_1 + m_2)$? In other words, the smaller the mass difference is, the slower the fall is.

Challenge 325, page 183: You should absolutely try to understand the origin of this expression. It allows understanding many important concepts of mechanics. The idea is that for small amplitudes, the acceleration of a pendulum of length l is due to gravity. Drawing a force diagram for a pendulum at a general angle α shows that

$$ma = -mg \sin \alpha$$

$$ml\frac{\text{d}^2\alpha}{\text{d}t^2} = -mg \sin \alpha$$

$$l\frac{\text{d}^2\alpha}{\text{d}t^2} = -g \sin \alpha \ . \tag{147}$$

For the mentioned small amplitudes (below 15°) we can approximate this to

$$l\frac{\text{d}^2\alpha}{\text{d}t^2} = -g\alpha \ . \tag{148}$$

This is the equation for a harmonic oscillation (i.e., a sinusoidal oscillation). The resulting motion is:

$$\alpha(t) = A \sin(\omega t + \varphi) \ . \tag{149}$$

The amplitude A and the phase φ depend on the initial conditions; however, the oscillation frequency is given by the length of the pendulum and the acceleration of gravity (check it!):

$$\omega = \sqrt{\frac{l}{g}} \ . \tag{150}$$

(For arbitrary amplitudes, the formula is much more complex; see the internet or special mechanics books for more details.)

Challenge 326, page 183: Walking speed is proportional to l/T, which makes it proportional to $l^{1/2}$. The relation is also true for animals in general. Indeed, measurements show that the maximum walking speed (thus not the running speed) across all animals is given by

$$v_{\text{maxwalking}} = (2.2 \pm 0.2)\,\text{m}^{1/2}/\text{s}\ \sqrt{l} \ . \tag{151}$$

Challenge 330, page 186: There is no obvious candidate formula. Can you find one?

Challenge 331, page 186: The acceleration due to gravity is $a = Gm/r^2 \approx 5\,\text{nm/s}^2$ for a mass of 75 kg. For a fly with mass $m_{\text{fly}} = 0.1\,\text{g}$ landing on a person with a speed of $v_{\text{fly}} = 1\,\text{cm/s}$

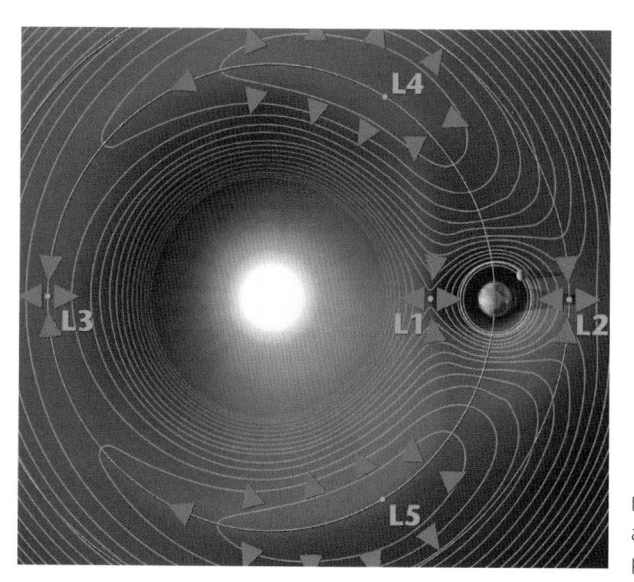

FIGURE 319 The Lagrangian points and the effective potential that produces them (NASA).

and deforming the skin (without energy loss) by $d = 0.3\,\text{mm}$, a person would be accelerated by $a = (v^2/d)(m_{\text{fly}}/m) = 0.4\,\mu\text{m/s}^2$. The energy loss of the inelastic collision reduces this value at least by a factor of ten.

Challenge 332, page 187: The calculation shows that a surprisingly high energy value is stored in thermal motion.

Challenge 333, page 188: Yes, the effect has been measured for skyscrapers. Can you estimate the values?

Challenge 336, page 188: The easiest way to see this is to picture gravity as a flux emanating from a sphere. This gives a $1/r^{d-1}$ dependence for the force and thus a $1/r^{d-2}$ dependence of the potential.

Challenge 338, page 190: Since the paths of free fall are ellipses, which are curves lying in a plane, this is obvious.

Challenge 340, page 191: A flash of light is sent to the Moon, where several Cat's-eyes have been deposited by the Lunokhod and Apollo missions. The measurement precision of the time a flash take to go and come back is sufficient to measure the Moon's distance change. For more details, see challenge 8.

Challenge 342, page 193: A body having zero momentum at spatial infinity is on a parabolic path. A body with a lower momentum is on an elliptic path and one with a higher momentum is on a hyperbolic path.

Challenge 345, page 194: The Lagrangian points L4 and L5 are on the orbit, 60° before and behind the orbiting body. They are stable if the mass ratio of the central and the orbiting body is sufficiently large (above 24.9).

Challenge 346, page 194: The Lagrangian point L3 is located on the orbit, but precisely on the other side of the central body. The Lagrangian point L1 is located on the line connecting the planet with the central body, whereas L2 lies outside the orbit, on the same line. If R is the radius of the orbit, the distance between the orbiting body and the L1 and L2 point is $\sqrt[3]{m/3M}\,R$, giving around 4 times the distance of the Moon for the Sun-Earth system. L1, L2 and L3 are saddle points, but effectively stable orbits exist around them. Many satellites make use of these properties, including

FIGURE 320 The famous 'vomit comet', a KC-135, performing a parabolic flight (NASA).

the famous WMAP satellite that measured the ripples of the big bang, which is located at the 'quiet' point L2, where the Sun, the Earth and the Moon are easily shielded and satellite temperature remains constant.

Challenge 347, page 197: This is a resonance effect, in the same way that a small vibration of a string can lead to large oscillation of the air and sound box in a guitar.

Challenge 349, page 199: The expression for the strength of tides, namely $2GM/d^3$, can be rewritten as $(8/3)\pi G\rho(R/d)^3$. Now, R/d is roughly the same for Sun and Moon, as every eclipse shows. So the density ρ must be much larger for the Moon. In fact, the ratio of the strengths (height) of the tides of Moon and Sun is roughly 7 : 3. This is also the ratio between the mass densities of the two bodies.

Challenge 350, page 200: The total angular momentum of the Earth and the Moon must remain constant.

Challenge 352, page 201: Wait for a solar eclipse.

Challenge 354, page 203: Unfortunately, the myth of 'passive gravitational mass' is spread by many books. Careful investigation shows that it is measured in exactly the same way as inertial mass.

Both masses are measured with the same machines and set-ups. And all these experiments mix and require both inertial and passive gravitational mass effects. For example, a balance or bathroom scale has to dampen out any oscillation, which requires inertial mass. Generally speaking, it seems impossible to distinguish inertial mass from the passive gravitational mass due to all the masses in the rest of the universe. In short, the two concepts are in fact identical.

Challenge 356, page 203: These problems occur because gravitational mass determines potential energy and inertial mass determines kinetic energy.

Challenge 358, page 205: Either they fell on inclined snowy mountain sides, or they fell into high trees, or other soft structures. The record was over 7 km of survived free fall. A recent case made the news in 2007 and is told in www.bbc.co.uk/jersey/content/articles/2006/12/20/michael_holmes_fall_feature.shtml.

Challenge 360, page 207: For a few thousand Euros, you can experience zero-gravity in a parabolic flight, such as the one shown in Figure 320. (Many 'photographs' of parabolic flights found on the internet are in fact computer graphics. What about this one?)

How does zero-gravity *feel*? It feels similar to floating under water, but without the resistance of the water. It also feels like the time in the air when one is diving into water. However, for cosmonauts, there is an additional feeling; when they rotate their head rapidly, the sensors for orientation in our ear are not reset by gravity. Therefore, for the first day or two, most cosmonauts have feelings of vertigo and of nausea, the so-called *space sickness*. After that time, the body adapts and the cosmonaut can enjoy the situation thoroughly.

Challenge 361, page 207: The centre of mass of a broom falls with the usual acceleration; the end thus falls faster.

Challenge 362, page 207: Just use energy conservation for the two masses of the jumper and the string. For more details, including the comparison of experimental measurements and theory, see N. Dubelaar & R. Brantjes, *De valversnelling bij bungee-jumping*, Nederlands tijdschrift voor natuurkunde 69, pp. 316–318, October 2003.

Challenge 363, page 207: About 1 ton.

Challenge 364, page 207: About 5 g.

Challenge 365, page 208: Your weight is roughly constant; thus the Earth must be round. On a flat Earth, the weight would change from place to place, depending on your distance from the border.

Challenge 366, page 208: Nobody ever claimed that the centre of mass is the same as the centre of gravity! The attraction of the Moon is negligible on the surface of the Earth.

Challenge 368, page 209: That is the mass of the Earth. Just turn the table on its head.

Challenge 370, page 209: The Moon will be about 1.25 times as far as it is now. The Sun then will slow down the Earth–Moon system rotation, this time due to the much smaller tidal friction from the Sun's deformation. As a result, the Moon will return to smaller and smaller distances to Earth. However, the Sun will have become a red giant by then, after having swallowed both the Earth and the Moon.

Challenge 372, page 210: As Galileo determined, for a swing (half a period) the ratio is $\sqrt{2}/\pi$. (See challenge 325). But not more than two, maybe three decimals of π can be determined in this way.

Challenge 373, page 210: Momentum conservation is not a hindrance, as any tennis racket has the same effect on the tennis ball.

Challenge 374, page 210: In fact, in velocity space, elliptic, parabolic and hyperbolic motions are all described by circles. In all cases, the hodograph is a circle.

Ref. 162

Challenge 375, page 211: This question is old (it was already asked in Newton's times) and deep. One reason is that stars are kept apart by rotation around the galaxy. The other is that galaxies are kept apart by the momentum they got in the big bang. Without the big bang, all stars would have collapsed together. In this sense, the big bang can be deduced from the attraction of gravitation and the immobile sky at night. We shall find out later that the darkness of the night sky gives a second argument for the big bang.

Challenge 376, page 211: The choice is clear once you notice that there is no section of the orbit which is concave towards the Sun. Can you show this?

Challenge 378, page 212: The escape velocity, from Earth, to leave the Solar System – without help of the other planets – is 42 km/s. However, if help by the other planets is allowed, it can be less than half that value (why?).

If the escape velocity from a body were the speed of light, the body would be a black hole; not even light could escape. Black holes are discussed in detail in the volume on relativity.

Vol. II, page 262

Challenge 379, page 212: Using a maximal jumping height of $h = 0.5$ m on Earth and an estimated asteroid density of $\rho = 3$ Mg/m^3, we get a maximum radius of $R^2 = 3gh/4\pi G\rho$, or $R \approx 2.4$ km.

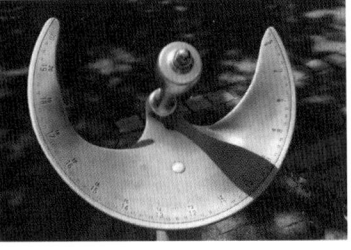

FIGURE 321 The analemma photographed, at local noon, from January to December 2002, at the Parthenon on Athen's Acropolis, and a precision sundial (© Anthony Ayiomamitis, Stefan Pietrzik).

Challenge 380, page 212: A handle of two bodies.

Challenge 382, page 212: In what does this argument differ from the more common argument that in the expression $ma = gMm/R^2$, the left m is inertial and the right m is gravitational?

Challenge 384, page 212: What counts is *local* verticality; with respect to it, the river always flows downhill.

Challenge 385, page 212: The shape of an analemma at local noon is shown in Figure 321. The shape is known since over 2000 years! The shape of the analemma also illustrates why the earliest sunrise is not at the longest day of the year.

The vertical extension of the analemma in the figure is due to the obliquity, i.e., the tilt of the Earth's axis (it is twice 23.45°). The horizontal extension is due to the combination of the obliquity and of the ellipticity of the orbit around the Sun. Both effects lead to roughly equal changes of the position of the Sun at local noon during the course of the year. The asymmetrical position of the central crossing point is purely due to the ellipticity of the orbit. The shape of the analemma, sometimes shown on globes, is built into the shadow pole or the reading curve of precision sundials. Examples are the one shown above and the one shown on page 45. For more details, see B. M. OLIVER, *The shape of the analemma*, Sky & Telescope 44, pp. 20–22, 1972, and the correction of the figures at 44, p. 303, 1972,

Challenge 386, page 215: Capture of a fluid body is possible if it is split by tidal forces.

Challenge 387, page 216: The tunnel would be an elongated ellipse in the plane of the Equator, reaching from one point of the Equator to the point at the antipodes. The time of revolution would not change, compared to a non-rotating Earth. See A. J. SIMONSON, *Falling down a hole through the Earth*, Mathematics Magazine 77, pp. 171–188, June 2004.

Challenge 389, page 216: The centre of mass of the Solar System can be as far as twice the radius from the centre of the Sun; it thus can be outside the Sun.

Challenge 390, page 216: First, during northern summer time the Earth moves faster around the Sun than during northern winter time. Second, shallow Sun's orbits on the sky give longer days because of light from when the Sun is below the horizon.

Challenge 391, page 216: Apart from the visibility of the Moon, no effect of the Moon on humans has ever been detected. Gravitational effects – including tidal effects – electrical effects, magnetic effects and changes in cosmic rays are all swamped by other effects. Indeed the gravity of passing trucks, factory electromagnetic fields, the weather and solar activity changes have larger influences on humans than the Moon. The locking of the menstrual cycle to the moon phase is a visual effect.

Challenge 392, page 216: Distances were difficult to measure. It is easy to observe a planet that is before the Sun, but it is hard to check whether a planet is behind the Sun. Phases of Venus are also predicted by the geocentric system; but the phases it predicts do not match the ones that are observed. Only the phases deduced from the heliocentric system match the observed ones. Venus orbits the Sun.

Challenge 393, page 216: See the mentioned reference.

Challenge 394, page 217: True.

Challenge 395, page 217: For each pair of opposite shell elements (drawn in yellow), the two attractions compensate.

Challenge 396, page 218: There is no practical way; if the masses on the shell could move, along the surface (in the same way that charges can move in a metal) this might be possible, provided that enough mass is available.

Challenge 400, page 219: Yes, one could, and this has been thought of many times, including by Jules Verne. The necessary speed depends on the direction of the shot with respect of the rotation of the Earth.

Challenge 401, page 219: Never. The Moon points always towards the Earth. The Earth changes position a bit, due to the ellipticity of the Moon's orbit. Obviously, the Earth shows phases.

Challenge 403, page 219: There are no such bodies, as the chapter of general relativity will show.

Challenge 405, page 221: The oscillation is a purely sinusoidal, or harmonic oscillation, as the restoring force increases linearly with distance from the centre of the Earth. The period T for a homogeneous Earth is $T = 2\pi\sqrt{R^3/GM} = 84\,\text{min}$.

Challenge 406, page 221: The period is the same for all such tunnels and thus in particular it is the same as the 84 min period that is valid also for the pole to pole tunnel. See for example, R. H. ROMER, *The answer is forty-two – many mechanics problems, only one answer*, Physics Teacher 41, pp. 286–290, May 2003.

Challenge 407, page 223: If the Earth were *not* rotating, the most general path of a falling stone would be an ellipse whose centre is the centre of the Earth. For a rotating Earth, the ellipse precesses. Simoson speculates that the spirographics swirls in the Spirograph Nebula, found at antwrp.gsfc.nasa.gov/apod/ap021214.html, might be due to such an effect. A special case is a path starting vertically at the equator; in this case, the path is similar to the path of the Foucault pendulum, a pointed star with about 16 points at which the stone resurfaces around the Equator. Ref. 177

Challenge 408, page 223: There is no simple answer: the speed depends on the latitude and on other parameters. The internet also provides videos of solar eclipses seen from space, showing how the shadow moves over the surface of the Earth.

Challenge 409, page 224: The centrifugal force must be equal to the gravitational force. Call the constant linear mass density d and the unknown length l. Then we have $GMd \int_R^{R+l} \mathrm{d}r/r^2 = \omega^2 d \int_R^{R+l} r\,\mathrm{d}r$. This gives $GMdl/(R^2 + Rl) = (2Rl + l^2)\omega^2 d/2$, yielding $l = 0.14\,\text{Gm}$. For more on space elevators or lifts, see challenge 574.

Challenge 411, page 224: The inner rings must rotate faster than the outer rings. If the rings were solid, they would be torn apart. But this reasoning is true only if the rings are inside a certain limit, the so-called *Roche limit*. The Roche limit is that radius at which gravitational force F_g and tidal force F_t cancel on the surface of the satellite. For a satellite with mass m and radius r, orbiting a central mass M at distance d, we look at the forces on a small mass μ on its surface. We get the condition $Gm\mu/r^2 = 2GM\mu r/d^3$. A bit of algebra yields the approximate Roche limit value

$$d_\text{Roche} = R\left(2\frac{\rho_M}{\rho_m}\right)^{1/3}. \tag{152}$$

Below that distance from a central mass M, fluid satellites cannot exist. The calculation shown here is only an approximation; the actual Roche limit is about two times that value.

Challenge 414, page 228: The load is 5 times the load while standing. This explains why race horses regularly break their legs.

Challenge 415, page 228: At school, you are expected to answer that the weight is the same. This is a good approximation. But in fact the scale shows a slightly larger weight for the steadily running hourglass compared to the situation where the all the sand is at rest. Looking at the momentum flow explains the result in a simple way: the only issue that counts is the momentum of the sand in the upper chamber, all other effects being unimportant. That momentum slowly decreases during running. This requires a momentum flow from the scale: the effective weight increases. See also the experimental confirmation and its explanation by F. TUINSTRA & B. F. TUINSTRA, *The weight of an hourglass*, Europhysics News 41, pp. 25–28, March 2010, also available online.

Ref. 86

If we imagine a photon bouncing up and down in a box made of perfect mirrors, the ideas from the hourglass puzzle imply that the scale shows an increased weight compared to the situation without a photon. The weight increase is Eg/c^2, where E is the energy of the photon, $g = 9.81 \text{ m/s}^2$ and c is the speed of light. This story is told by E. HUGGINS, *Weighing photons using bathroom scales: a thought experiment*, The Physics Teacher 48, pp. 287–288, May 2010,

Challenge 416, page 229: The electricity consumption of a rising escalator indeed increases when the person on it walks upwards. By how much?

Challenge 417, page 229: Knowledge is power. Time is money. Now, power is defined as work per time. Inserting the previous equations and transforming them yields

$$\text{money} = \frac{\text{work}}{\text{knowledge}} , \qquad (153)$$

which shows that the less you know, the more money you make. That is why scientists have low salaries.

Challenge 418, page 229: In reality muscles keep an object above ground by continuously lifting and dropping it; that requires energy and work.

Challenge 421, page 234: Yes, because side wind increases the effective speed v in air due to vector addition, and because air resistance is (roughly) proportional to v^2.

Challenge 422, page 234: The lack of static friction would avoid that the fluid stays attached to the body; the so-called boundary layer would not exist. One then would have no wing effect.

Challenge 424, page 235: True?

Challenge 426, page 236: From $dv/dt = g - v^2(1/2c_w A\rho/m)$ and using the abbreviation $c = 1/2c_w A\rho$, we can solve for $v(t)$ by putting all terms containing the variable v on one side, all terms with t on the other, and integrating on both sides. We get $v(t) = \sqrt{gm/c} \tanh \sqrt{cg/m}\, t$.

Challenge 427, page 237: For extended deformable bodies, the intrinsic properties are given by the mass density – thus a function of space and time – and the state is described by the density of kinetic energy, local linear and angular momentum, as well as by its stress and strain distributions.

Challenge 428, page 237: Electric charge.

Challenge 429, page 238: The phase space has $3N$ position coordinates and $3N$ momentum coordinates.

Challenge 430, page 238: We recall that when a stone is thrown, the initial conditions summarize the effects of the thrower, his history, the way he got there etc.; in other words, initial conditions summarize the past of a system, i.e., the effects that the environment had during the history

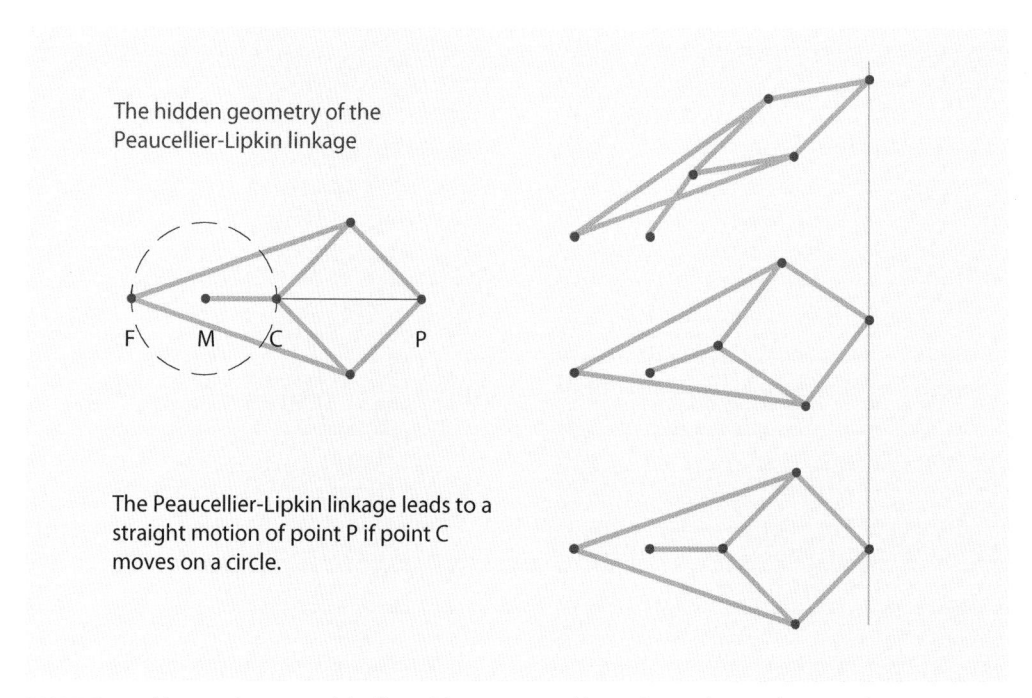

The hidden geometry of the
Peaucellier-Lipkin linkage

F M C   P

The Peaucellier-Lipkin linkage leads to a
straight motion of point P if point C
moves on a circle.

FIGURE 322 How to draw a straight line with a compass (drawn by Zach Joseph Espiritu).

of a system. Therefore, the universe has no initial conditions and no phase space. If you have found reasons to answer yes, you overlooked something. Just go into more details and check whether the concepts you used apply to the universe. Also define carefully what you mean by 'universe'.

Challenge 431, page 238: The light mill is an example.

Vol. III, page 122

Challenge 433, page 240: A system showing energy or matter motion faster than light would imply that for such systems there are observers for which the order between cause and effect are reversed. A space-time diagram (and a bit of exercise from the section on special relativity) shows this.

Challenge 434, page 240: If reproducibility would not exist, we would have difficulties in checking observations; also reading the clock is an observation. The connection between reproducibility and time shall become important in the final part of our adventure.

Challenge 435, page 241: Even if surprises were only rare, each surprise would make it impossible to define time just before and just after it.

Challenge 438, page 242: Of course; moral laws are summaries of what others think or will do about personal actions.

Challenge 439, page 243: The fastest glide path between two points, the *brachistochrone*, turns out to be the *cycloid*, the curve generated by a point on a wheel that is rolling along a horizontal plane.

The proof can be found in many ways. The simplest is by Johann Bernoulli and is given on en.wikipedia.org/wiki/Brachistochrone_problem.

Challenge 441, page 244: When F, C and P are aligned, this circle has a radius given by $R = \sqrt{\text{FCFP}}$; F is its centre. In other words, the Peaucellier-Lipkin linkage realizes an inversion at a circle.

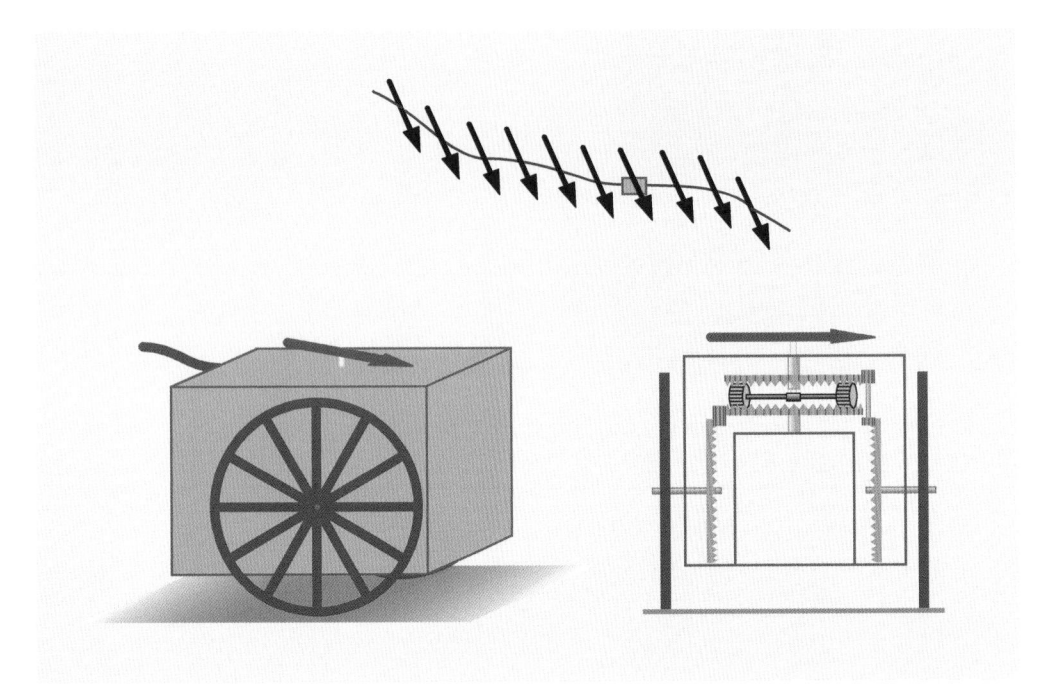

FIGURE 323 The mechanism inside the south-pointing carriage.

FIGURE 324 Falling brick chimneys – thus with limited stiffness – fall with a V shape (© John Glaser, Frank Siebner).

Challenge 442, page 244: When F, C and P are aligned, the circle to be followed has a radius given by half the distance FC; its centre lies midway between F and C. Figure 322 illustrates the situation.

Challenge 443, page 244: Figure 323 shows the most credible reconstruction of a south-pointing carriage.

Challenge 445, page 245: The water is drawn up along the sides of the spinning egg. The fastest way to empty a bottle of water is to spin the water while emptying it.

Challenge 446, page 246: The right way is the one where the chimney falls like a V, not like an inverted V. See challenge 361 on falling brooms for inspiration on how to deduce the answer.

Two examples are shown in Figure 324. It turns out that the chimney breaks (if it is not fastened to the base) at a height between half or two thirds of the total, depending at the angle at which this happens. For a complete solution of the problem, see the excellent paper G. VARESCHI & K. KAMIYA, *Toy models for the falling chimney*, American Journal of Physics 71, pp. 1025–1031, 2003.

Challenge 448, page 252: The definition of the integral given in the text is a simplified version of the so-called *Riemann integral*. It is sufficient for all uses in nature. Have a look at its exact definition in a mathematics text if you want more details.

Challenge 454, page 255: In one dimension, the expression $F = ma$ can be written as $-dV/dx = md^2x/dt^2$. This can be rewritten as $d(-V)/dx - d/dt[d/d\dot{x}(\frac{1}{2}m\dot{x}^2)] = 0$. This can be expanded to $\partial/\partial x(\frac{1}{2}m\dot{x}^2 - V(x)) - d/[\partial/\partial\dot{x}(\frac{1}{2}m\dot{x}^2 - V(x))] = 0$, which is Lagrange's equation for this case.

Challenge 456, page 256: Do not despair. Up to now, nobody has been able to imagine a universe (that is not necessarily the same as a 'world') different from the one we know. So far, such attempts have always led to logical inconsistencies.

Challenge 458, page 256: The two are equivalent since the equations of motion follow from the principle of minimum action and at the same time the principle of minimum action follows from the equations of motion.

Challenge 460, page 258: For gravity, all three systems exist: rotation in galaxies, pressure in planets and the Pauli pressure in stars that is due to Pauli's exclusion principle. Against the strong interaction, the exclusion principle acts in nuclei and neutron stars; in neutron stars maybe also rotation and pressure complement the Pauli pressure. But for the electromagnetic interaction there are no composites other than our everyday matter, which is organized by the Pauli's exclusion principle alone, acting among electrons.

Vol. IV, page 135

Challenge 461, page 259: Aggregates often form by matter converging to a centre. If there is only a small asymmetry in this convergence – due to some external influence – the result is a final aggregate that rotates.

Challenge 462, page 262: Angular momentum is the change with respect to angle, whereas rotational energy is again the change with respect to time, as all energy is.

Challenge 463, page 262: Not in this way. A small change can have a large effect, as every switch shows. But a small change in the brain must be communicated outside, and that will happen roughly with a $1/r^2$ dependence. That makes the effects so small, that even with the most sensitive switches – which for thoughts do not exist anyway – no effects can be realized.

Challenge 465, page 262: This is a wrong question. $T - U$ is not minimal, only its average is.

Challenge 466, page 263: No. A system tends to a minimum potential only if it is dissipative. One could, however, deduce that conservative systems oscillate around potential minima.

Challenge 467, page 263: The relation is

$$\frac{c_1}{c_2} = \frac{\sin\alpha_1}{\sin\alpha_2} \ . \tag{154}$$

The particular speed ratio between air (or vacuum, which is almost the same) and a material gives the *index of refraction n*:

$$n = \frac{c_1}{c_0} = \frac{\sin\alpha_1}{\sin\alpha_0} \tag{155}$$

Challenge 468, page 263: The principle for the growth of trees is simply the minimum of potential energy, since the kinetic energy is negligible. The growth of vessels inside animal bodies is minimized for transport energy; that is again a minimum principle. The refraction of light is the path of shortest time; thus it minimizes change as well, if we imagine light as moving entities moving without any potential energy involved.

Challenge 469, page 264: Special relativity requires that an invariant measure of the action exist. It is presented later in the walk.

Vol. VI, page 106 **Challenge 470**, page 264: The universe is not a physical system. This issue will be discussed in detail later on.

Challenge 471, page 264: Use either the substitution $u = \tan t/2$ or use the historical trick

$$\sec\varphi = \tfrac{1}{2}\left(\frac{\cos\varphi}{1+\sin\varphi} + \frac{\cos\varphi}{1-\sin\varphi}\right). \tag{156}$$

Challenge 472, page 264: A skateboarder in a cycloid has the same oscillation time independently of the oscillation amplitude. But a half-pipe needs to have vertical ends, in order to avoid jumping outside it. A cycloid never has a vertical end.

Challenge 475, page 267: We talk to a person because we know that somebody understands us. Thus we assume that she somehow sees the same things we do. That means that observation is partly viewpoint-independent. Thus nature is symmetric.

Challenge 476, page 270: Memory works because we recognize situations. This is possible because situations over time are similar. Memory would not have evolved without this reproducibility.

Challenge 477, page 271: Taste differences are not fundamental, but due to different viewpoints and – mainly – to different experiences of the observers. The same holds for feelings and judgements, as every psychologist will confirm.

Challenge 478, page 273: The integers under addition form a group. Does a painter's set of oil colours with the operation of mixing form a group?

Challenge 480, page 273: There is only one symmetry operation: a rotation about π around the central point. That is the reason that later on the group D_4 is only called the approximate symmetry group of Figure 202.

Challenge 486, page 278: Scalar is the magnitude of any vector; thus the speed, defined as $v = |v|$, is a scalar, whereas the velocity v is not. Thus the length of any vector (or pseudo-vector), such as force, acceleration, magnetic field, or electric field, is a scalar, whereas the vector itself is not a scalar.

Challenge 489, page 278: The charge distribution of an extended body can be seen as a sum of a point charge, a charge dipole, a charge quadrupole, a charge octupole, etc. The quadrupole is described by a tensor.

Compare: The inertia against motion of an extended body can be seen as sum of a point mass, a mass dipole, a mass quadrupole, a mass octupole, etc. The mass quadrupole is described by the moment of inertia.

Challenge 493, page 281: The conserved charge for rotation invariance is angular momentum.

Challenge 497, page 285: The graph is a *logarithmic spiral* (can you show this?); it is illustrated in Figure 325. The travelled distance has a simple answer.

Challenge 498, page 285: An oscillation has a period in time, i.e., a discrete time translation symmetry. A wave has both discrete time and discrete space translation symmetry.

Challenge 499, page 285: Motion reversal is a symmetry for any closed system; despite the observations of daily life, the statements of thermodynamics and the opinion of several famous physicists (who form a minority though) all ideally closed systems are reversible.

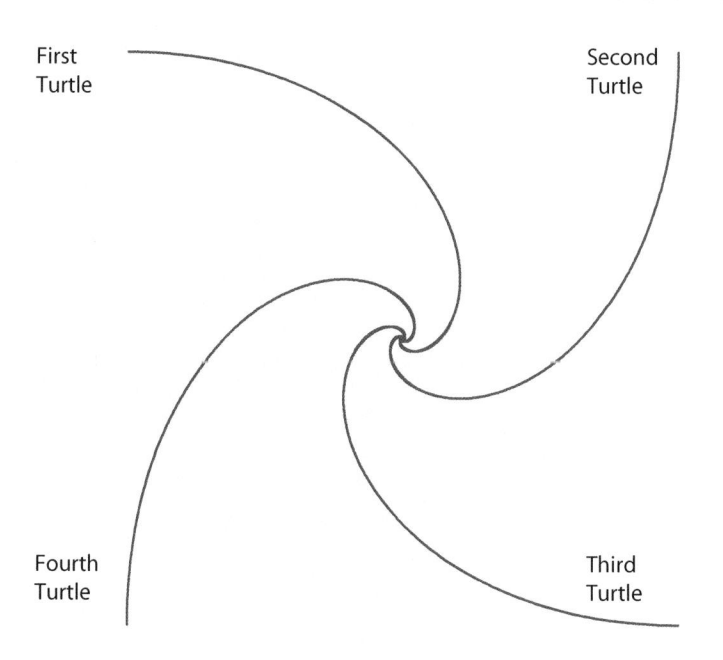

FIGURE 325 The motion of four turtles chasing each other (drawn by Zach Joseph Espiritu).

Challenge 500, page 285: The symmetry group is a Lie group and called U(1), for 'unitary group in 1 dimension'.

Challenge 501, page 285: See challenge 301

Challenge 502, page 285: There is no such thing as a 'perfect' symmetry.

Challenge 504, page 286: The rotating telephone dial had the digits 1 to 0 on the corners of a regular 14-gon. The even and the odd numbers were on the angles of regular heptagons.

Challenge 508, page 289: Just insert $x(t)$ into the Lagrangian $L = 0$, the minimum possible value for a system that transforms all kinetic energy into potential energy and vice versa.

Challenge 517, page 300: The potential energy is due to the 'bending' of the medium; a simple displacement produces no bending and thus contains no energy. Only the gradient captures the bending idea.

Challenge 519, page 301: The phase changes by π.

Challenge 520, page 301: A wave that carries angular momentum has to be transversal and has to propagate in three dimensions.

Challenge 521, page 301: Waves can be damped to extremely low intensities. If this is not possible, the observation is not a wave.

Challenge 522, page 303: The way to observe diffraction and interference with your naked fingers is told on page 101 in volume III.

Challenge 533, page 315: Interference can make radio signals unintelligible. Due to diffraction, radio signals are weakened behind a wall; this is valid especially for short wavelengths, such as those used in mobile phones. Refraction makes radio communication with submarines impossible for usual radio frequencies. Dispersion in glass fibres makes it necessary to add repeaters in sea-cables roughly every 100 km. Damping makes it impossible hear somebody speaking

CHALLENGE HINTS AND SOLUTIONS

at larger distances. Radio signals can loose their polarisation and thus become hard to detect by usual Yagi antennas that have a fixed polarisation.

Challenge 535, page 320: Skiers scrape snow from the lower side of each bump towards the upper side of the next bump. This leads to an upward motion of ski bumps.

Challenge 536, page 320: If the distances to the loudspeaker is a few metres, and the distance to the orchestra is 20 m, as for people with enough money, the listener at home hears it first.

Challenge 537, page 320: As long as the amplitude is small compared to the length l, the period T is given by

$$T = 2\pi\sqrt{\frac{l}{g}} \ . \tag{157}$$

The formula does not contain the mass m at all. Independently of the mass m at its end, the pendulum has always the same period. In particular, for a length of 1 m, the period is about 2 s. Half a period, or one swing thus takes about 1 s. (This is the original reason for choosing the unit of metre.)

For an extremely long pendulum, the answer is a finite value though, and corresponds to the situation of challenge 26.

Challenge 538, page 320: In general, the body moves along an ellipse (as for planets around the Sun) but with the fixed point as centre. In contrast to planets, where the Sun is in a *focus* of the ellipse and there is a perihelion and an aphelion, such a body moves *symmetrically* around the *centre* of the ellipse. In special cases, the body moves back and forward along a straight segment.

Challenge 540, page 320: This follows from the formula that the frequency of a string is given by $f = \sqrt{T/\mu}/(2l)$, where T is the tension, μ is the linear mass density, and l is the length of a string. This is discussed in the beautiful paper by G. BARNES, *Physics and size in biological systems*, The Physics Teacher 27, pp. 234–253, 1989.

Challenge 542, page 321: The sound of thunder or of car traffic gets lower and lower in frequency with increasing distance.

Challenge 545, page 321: Neither; both possibilities are against the properties of water: in surface waves, the water molecules move in circles.

Challenge 546, page 322: Swimmers are able to cover 100 m in 48 s, or slightly better than 2 m/s. (Swimmer with fins achieve just over 3 m/s.) With a body length of about 1.9 m, the critical speed is 1.7 m/s. That is why short distance swimming depends on training; for longer distances the technique plays a larger role, as the critical speed has not been attained yet. The formula also predicts that on the 1500 m distance, a 2 m tall swimmer has a potential advantage of over 45 s on one with body height of 1.8 m. In addition, longer swimmers have an additional advantage: they swim shorter distances in pools (why?). It is thus predicted that successful long-distance swimmers will get taller and taller over time. This is a pity for a sport that so far could claim to have had champions of all sizes and body shapes, in contrast to many other sports.

Challenge 549, page 324: To reduce noise reflection and thus hall effects. They effectively diffuse the arriving wave fronts.

Challenge 551, page 324: Waves in a river are never elliptical; they remain circular.

Challenge 552, page 324: The lens is a cushion of material that is 'transparent' to sound. The speed of sound is faster in the cushion than in the air, in contrast to a glass lens, where the speed of light is slower in the glass. The shape is thus different: the cushion must look like a large biconcave optical lens.

Challenge 553, page 324: Experiments show that the sound does not depend on air flows (find out how), but does depend on external sound being present. The sound is due to the selective amplification by the resonances resulting from the geometry of the shell shape.

Challenge 554, page 324: The Sun is always at a different position than the one we observe it to be. What is the difference, measured in angular diameters of the Sun? Despite this position difference, the timing of the sunrise is determoned by the position of the horizon, not by the position of the Sun. (Imagine the it would not: in that case a room would not get dark when the window is closed, but eight minutes later ...) In short, there is no measurable effect of the speed of light on the sunrise.

Challenge 557, page 326: An overview of systems being tested at present can be found in K. - U. GRAW, *Energiereservoir Ozean*, Physik in unserer Zeit 33, pp. 82–88, Februar 2002. See also *Oceans of electricity – new technologies convert the motion of waves into watts*, Science News 159, pp. 234–236, April 2001.

Challenge 558, page 326: In everyday life, the assumption is usually justified, since each spot can be approximately represented by an atom, and atoms can be followed. The assumption is questionable in situations such as turbulence, where not all spots can be assigned to atoms, and most of all, in the case of motion of the vacuum itself. In other words, for gravity waves, and in particular for the quantum theory of gravity waves, the assumption is not justified.

Challenge 564, page 333: There are many. One would be that the transmission and thus reflection coefficient for waves would almost be independent of wavelength.

Challenge 565, page 334: A drop with a diameter of 3 mm would cover a surface of $7.1 \, \text{m}^2$ with a 2 nm film.

Challenge 566, page 338: The wind will break tall trees that are too thin. For small and thus thin trees, the wind does not damage.

Challenge 567, page 338: The critical height for a column of material is given by $h_{\text{crit}}^4 = \frac{\beta}{4\pi g} m \frac{E}{\rho^2}$, where $\beta \approx 1.9$ is the constant determined by the calculation when a column buckles under its own weight.

Challenge 569, page 339: One possibility is to describe particles as clouds; another is given in the last part of the text.

Challenge 570, page 341: The results gives a range between 1 and $8 \cdot 10^{23}$.

Ref. 257

Challenge 572, page 345: Check your answers with the delightful text by P. GOLDRICH, S. MAHAJAN & S. PHINNEY, *Order-of-Magnitude Physics: Understanding the World with Dimensional Analysis, Educated Guesswork, and White Lies*, available on the internet.

Challenge 573, page 345: Glass shatters, glass is elastic, glass shows transverse sound waves, glass does not flow (in contrast to what many books state), not even on scale of centuries, glass molecules are fixed in space, glass is crystalline at small distances, a glass pane supported at the ends does not hang through.

Challenge 574, page 345: No metal wire allows building such a long wire or rope. Only the idea of carbon nanotubes has raised the hope again; some dream of wire material based on them, stronger than any material known so far. However, no such material is known yet. The system faces many dangers, such as fabrication defects, lightning, storms, meteoroids and space debris. All would lead to the breaking of the wires – if such wires will ever exist. But the biggest of all dangers is the lack of cash to build it. Nevertheless, numerous people are working towards the goal.

Challenge 575, page 346: The $3 \times 3 \times 3$ cube has a rigid system of three perpendicular axes, on which a square can rotate at each of the 6 ends. The other squares are attaches to pieces moving around theses axes. The $4 \times 4 \times 4$ cube is different though; just find out. From $7 \times 7 \times 7$ onwards, the parts do not all have the same size or shape. The present limit on the segment number in commercially available 'cubes' is $17 \times 17 \times 17$! It can be found at www.shapeways.com/shops/

oskarpuzzles. The website www.oinkleburger.com/Cube/applet allows playing with virtual cubes up to $100 \times 100 \times 100$, and more.

Challenge 578, page 347: A medium-large earthquake would be generated.

Challenge 579, page 347: A stalactite contains a thin channel along its axis through which the water flows, whereas a stalagmite is massive throughout.

Challenge 580, page 347: About 1 part in a thousand.

Challenge 581, page 348: Even though the iron core of the Earth formed by collecting the iron from colliding asteroids which then sunk into the centre of the Earth, the scheme will not work today: in its youth, the Earth was much more liquid than today. The iron will most probably not sink. In addition, there is no known way to build a measurement probe that can send strong enough sound waves for this scheme. The temperature resistance is also an issue, but this may be solvable.

Challenge 583, page 350: Atoms are not infinitely hard, as quantum theory shows. Atoms are more similar to deformable clouds.

Vol. IV, page 80

Challenge 586, page 360: If there is no friction, all three methods work equally fast – including the rightmost one.

Challenge 589, page 362: The constant k follows from the conservation of energy and that of mass:

$$k = \sqrt{\frac{2}{\rho(A_1^2/A_2^2 - 1)}} \; . \tag{158}$$

The cross sections are denoted by A and the subscript 1 refers to any point far from the constriction, and the subscript 2 to the constriction.

Challenge 592, page 369: The pressure destroys the lung. Snorkeling is only possible at the water surface, not below the water! This experiment is even dangerous when tried in your own bathtub! Breathing with a long tube is only possible if a pump at the surface pumps air down the tube at the correct pressure.

Challenge 594, page 369: Some people notice that in some cases friction is too high, and start sucking at one end of the tube to get the flow started; while doing so, they can inhale or swallow gasoline, which is poisonous.

Challenge 599, page 372: Calculation yields $N = J/j = (0.0001\,\text{m}^3/\text{s})/(7\,\mu\text{m}^2 0.0005\,\text{m/s})$, or about $6 \cdot 10^9$; in reality, the number is much larger, as most capillaries are closed at a given instant. The reddening of the face shows what happens when all small blood vessels are opened at the same time.

Challenge 600, page 373: Throwing the stone makes the level fall, throwing the water or the piece of wood leaves it unchanged.

Challenge 601, page 373: The ship rises higher into the sky. (Why?)

Challenge 603, page 373: The motion of a helium-filled balloon is opposite to that of an air-filled balloon or of people: the helium balloon moves towards the front when the car accelerates and to the back when the car decelerates. It also behaves differently in bends. Several films on the internet show the details.

Challenge 606, page 373: The pumps worked in suction; but air pressure only allows $10\,\text{m}$ of height difference for such systems.

Challenge 607, page 373: This argument is comprehensible only when we remember that 'twice the amount' means 'twice as many molecules'.

Challenge 608, page 373: The alcohol is frozen and the chocolate is put around it.

TABLE 62 *Gaseous* composition of *dry* air, at *present* time[a] (sources: NASA, IPCC).

GAS	SYMBOL	VOLUME PART[b]
Nitrogen	N_2	78.084 %
Oxygen (pollution dependent)	O_2	20.946 %
Argon	Ar	0.934 %
Carbon dioxide (in large part due to human pollution)	CO_2	403 ppm
Neon	Ne	18.18 ppm
Helium	He	5.24 ppm
Methane (mostly due to human pollution)	CH_4	1.79 ppm
Krypton	Kr	1.14 ppm
Hydrogen	H_2	0.55 ppm
Nitrous oxide (mostly due to human pollution)	N_2O	0.3 ppm
Carbon monoxide (partly due to human pollution)	CO	0.1 ppm
Xenon	Xe	0.087 ppm
Ozone (strongly influenced by human pollution)	O_3	0 to 0.07 ppm
Nitrogen dioxide (mostly due to human pollution)	NO_2	0.02 ppm
Iodine	I_2	0.01 ppm
Ammonia (mostly due to human pollution)	NH_3	traces
Radon	Ra	traces
Halocarbons and other fluorine compounds (all being humans pollutants)	20 types	0.0012 ppm
Mercury, other metals, sulfur compounds, other organic compounds (all being human pollutants)	numerous	concentration varies

a. Wet air can contain up to 4 % water vapour, depending on the weather. *Apart from gases,* air can contain water droplets, ice, sand, dust, pollen, spores, volcanic ash, forest fire ash, fuel ash, smoke particles, pollutants of all kinds, meteoroids and cosmic ray particles. *During the history* of the Earth, the gaseous composition varied strongly. In particular, oxygen is part of the atmosphere only in the second half of the Earth's lifetime.
b. The abbreviation *ppm* means 'parts per million'.

Challenge 609, page 374: The author suggested in an old edition of this text that a machine should be based on the same machines that throw the clay pigeons used in the sports of trap shooting and skeet. In the meantime, Lydéric Bocquet and Christophe Clanet have built such a stone-skipping machine, but using a different design; a picture can be found on the website ilm-perso.univ-lyon1.fr/~lbocquet.

Challenge 610, page 374: The third component of *air* is the noble gas argon, making up about 1 %. A longer list of components is given in Table 62.

Challenge 611, page 374: The *pleural cavity* between the lungs and the thorax is permanently below atmospheric pressure, usually 5 mbar, but even 10 mbar at inspiration. A hole in it, formed for example by a bullet, a sword or an accident, leads to the collapse of the lung – the so-called *pneumothorax* – and often to death. Open chest operations on people have became possible only after the surgeon Ferdinand Sauerbruch learned in 1904 how to cope with the problem. Nowadays however, surgeons keep the lung under *higher* than atmospheric pressure until

FIGURE 326 A way to ride head-on against the wind using wind power (© Tobias Klaus).

everything is sealed again.

Challenge 612, page 374: The fountain shown in the figure is started by pouring water into the uppermost container. The fountain then uses the air pressure created by the water flowing downwards.

Challenge 613, page 374: Yes. The bulb will not resist two such cars though.

Challenge 614, page 375: Radon is about 8 times as heavy as air; it is he densest gas known. In comparison, Ni(CO) is 6 times, $SiCl_4$ 4 times heavier than air. Mercury vapour (obviously also a gas) is 7 times heavier than air. In comparison, bromine vapour is 5.5 times heavier than air.

Challenge 616, page 375: Yes, as the *ventomobil* shown in Figure 326 proves. It achieves the feat already for low wind speeds.

Challenge 617, page 376: None.

Challenge 619, page 376: He brought the ropes into the cabin by passing them through liquid mercury.

Challenge 621, page 376: There are no official solutions for these questions; just check your assumptions and calculations carefully. The internet is full of such calculations.

Challenge 622, page 377: The soap flows down the bulb, making it thicker at the bottom and thinner at the top, until it reaches the thickness of two molecular layers. Later, it bursts.

Challenge 623, page 377: The temperature leads to evaporation of the involved liquid, and the vapour prevents the direct contact between the two non-gaseous bodies.

Challenge 624, page 377: For this to happen, friction would have to exist on the microscopic scale and energy would have to disappear.

Challenge 625, page 378: The longer funnel is empty before the short one. (If you do not believe it, try it out.) In the case that the amount of water in the funnel outlet can be neglected, one can use energy conservation for the fluid motion. This yields the famous Bernoulli equation $p/\rho + gh + v^2/2 = $ const, where p is pressure, ρ the density of water, and g is 9.81 m/s^2. Therefore, the speed v is higher for greater lengths h of the thin, straight part of the funnel: the longer funnel empties first.

But this is strange: the formula gives a simple free fall relation, as the *air* pressure is the same above and below and disappears from the calculation. The expression for the speed is thus independent of whether a tube is present or not. The real reason for the faster emptying of the tube is thus that a tube forces more water to flow out than the lack of a tube. Without tube, the diameter of the water flow *diminishes* during fall. With tube, it stays *constant*. This difference leads to the faster emptying for longer tubes.

Alternatively, you can look at the *water* pressure value *inside* the funnel. You will discover that the water pressure is lowest at the start of the exit tube. This internal water pressure is lower for longer tubes and sucks out the water faster in those cases.

Challenge 626, page 378: The eyes of fish are positioned in such a way that the pressure reduction by the flow is compensated by the pressure increase of the stall. By the way, their heart is positioned in such a way that it is helped by the underpressure.

Challenge 628, page 378: This feat has been achieved for lower mountains, such as the Monte Bianco in the Alps. At present however, there is no way to safely hover at the high altitudes of the Himalayas.

Challenge 630, page 378: Press the handkerchief in the glass, and lower the glass into the water with the opening first, while keeping the opening horizontal. This method is also used to lower people below the sea. The paper ball in the bottle will fly towards you. Blowing into a funnel will keep the ping-pong ball tightly into place, and the more so the stronger you blow. Blowing through a funnel towards a candle will make it lean towards you.

Challenge 637, page 388: In 5000 million years, the present method will stop, and the Sun will become a red giant. But it will burn for many more years after that.

Challenge 638, page 390: Bernoulli argued that the temperature describes the average kinetic energy of the constituents of the gas. From the kinetic energy he deduced the average momentum of the constituents. An average momentum leads to a pressure. Adding the details leads to the ideal gas relation.

Challenge 639, page 390: The answer depends on the size of the balloons, as the pressure is not a monotonous function of the size. If the smaller balloon is not too small, the smaller balloon wins.

Challenge 642, page 391: Measure the area of contact between tires and street (all four) and then multiply by 200 kPa, the usual tire pressure. You get the weight of the car.

Challenge 646, page 394: If the average square displacement is proportional to time, the liquid is made of smallest particles. This was confirmed by the experiments of Jean Perrin. The next step is to deduce the number of these particles from the proportionality constant. This constant, defined by $\langle d^2 \rangle = 4Dt$, is called the diffusion constant (the factor 4 is valid for random motion in two dimensions). The diffusion constant can be determined by watching the motion of a particle under the microscope.

We study a Brownian particle of radius a. In two dimensions, its square displacement is given by

$$\langle d^2 = \rangle \frac{4kT}{\mu} t \, , \tag{159}$$

where k is the Boltzmann constant and T the temperature. The relation is deduced by studying the motion of a particle with drag force $-\mu v$ that is subject to random hits. The linear drag coefficient μ of a sphere of radius a is given by

$$\mu = 6\pi\eta a \, , \tag{160}$$

where η is the kinematic viscosity. In other words, one has

$$k = \frac{6\pi\eta a}{4T} \frac{\langle d^2 \rangle}{t} \, . \tag{161}$$

All quantities on the right can be measured, thus allowing us to determine the Boltzmann constant k. Since the ideal gas relation shows that the ideal gas constant R is related to the Boltzmann constant by $R = N_A k$, the Avogadro constant N_A that gives the number of molecules in a mole is also found in this way.

Challenge 651, page 402: The possibility of motion inversion for all observed phenomena is indeed a fundamental property of nature. It has been confirmed for all interactions and all experiments every performed. Independent of this is the fact that realizing the inversion might be extremely hard, because inverting the motion of many atoms is usually not feasible.

Challenge 652, page 403: This is a trick question. To a good approximation, any tight box is an example. However, if we ask for complete precision, all systems radiate some energy, loose some atoms or particles and bend space; *ideal* closed systems do *not* exist.

Vol. VI, page 106 **Challenge 657**, page 405: We will find out later that the universe is not a physical system; thus the concept of entropy does not apply to it. Thus the universe is neither isolated nor closed.

Challenge 659, page 406: Egg white starts to harden at lower temperature than yolk, but for complete hardening, the opposite is true. White hardens *completely* at 80°C, egg yolk hardens considerably at 66 to 68°C. Cook an egg at the latter temperature, and the feat is possible; the white remains runny, but does not remain transparent, though. Note again that the cooking time plays no role, only the precise temperature value.

Challenge 661, page 407: Yes, the effect is easily noticeable.

Challenge 664, page 407: Hot air is less dense and thus wants to rise.

Challenge 665, page 407: Keep the paper wet.

Challenge 666, page 407: Melting ice at 0°C to water at 0°C takes 334 kJ/kg. Cooling water by 1°C or 1 K yields 4.186 kJ/kgK. So the hot water needs to cool down to 20.2°C to melt the ice, so that the final mixing temperature will be 10.1°C.

Challenge 667, page 408: The air had to be dry.

Challenge 668, page 408: In general, it is impossible to draw a line through three points. Since absolute zero and the triple point of water are fixed in magnitude, it was practically a sure bet that the boiling point would not be at precisely 100°C.

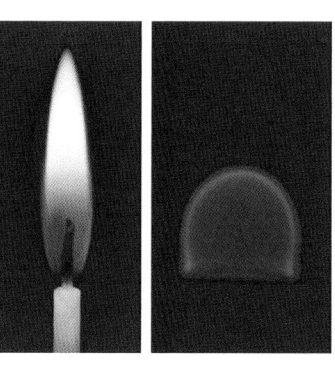

FIGURE 327 A candle on Earth and in microgravity (NASA).

Challenge 669, page 408: No, as a water molecule is heavier than that. However, if the water is allowed to be dirty, it is possible. What happens if the quantum of action is taken into account?

Challenge 670, page 408: The danger is not due to the amount of energy, but due to the time in which it is available.

Challenge 671, page 409: The internet is full of solutions.

Challenge 672, page 409: There are 2^n possible sequences of n coin throws. Of those, $n!/(\frac{n}{2}!)^2$ contain $n/2$ heads and $n/2$ tails. For a fair coin, the probability p of getting $n/2$ heads in n throws is thus

$$p = \frac{n!}{2^n \left(\frac{n}{2}!\right)^2} \ . \tag{162}$$

We approximate this result with the help of Gosper's formula $n! \approx \sqrt{(2n + \frac{1}{3})\pi} \ (\frac{n}{e})^n$ and get

$$p \approx \frac{\sqrt{(2n + \frac{1}{3})\pi} \ (\frac{n}{e})^n}{2^n \left(\sqrt{(n + \frac{1}{3})\pi} \ (\frac{n}{2e})^{\frac{n}{2}}\right)^2} = \frac{\sqrt{2n + \frac{1}{3}}}{(n + \frac{1}{3})\sqrt{\pi}} \ . \tag{163}$$

For $n = 1\,000\,000$, we get a probability $p \approx 0.0007979$, thus a rather small value between $\frac{1}{1254}$ and $\frac{1}{1253}$.

Challenge 673, page 409: The entropy can be defined for the universe as a whole only if the universe is a closed system. But is the universe closed? Is it a system? This issue is discussed in the final part of our adventure.

Challenge 676, page 410: For such small animals the body temperature would fall too low. They could not eat fast enough to get the energy needed to keep themselves warm.

Challenge 679, page 410: The answer depends on the volume, of course. But several families have died overnight because they had modified their mobile homes to be airtight.

Challenge 680, page 410: The metal salts in the ash act as catalysts, and the sugar burns instead of just melting. Watch the video of the experiment at www.youtube.com/watch?v=BfBgAaeaVgk.

Challenge 685, page 411: It is about 10^{-9} that of the Earth.

Challenge 687, page 411: The thickness of the folds in the brain, the bubbles in the lung, the density of blood vessels and the size of biological cells.

Challenge 688, page 411: The mercury vapour above the liquid gets saturated.

Challenge 689, page 411: A dedicated NASA project studied this question. Figure 327 gives an example comparison. You can find more details on their website.

Challenge 690, page 411: The risks due to storms and the financial risks are too high.

Challenge 691, page 412: The vortex inside the tube is cold near its axis and hot in the regions away from the axis. Through the membrane in the middle of the tube (shown in Figure 285 on page 412) the air from the axis region is sent to one end and the air from the outside region to the other end. The heating of the outside region is due to the work that the air rotating inside has to do on the air outside to get a rotation that consumes angular momentum. For a detailed explanation, see the beautiful text by MARK P. SILVERMAN, *And Yet it Moves: Strange Systems and Subtle Questions in Physics*, Cambridge University Press, 1993, p. 221.

Challenge 692, page 412: No.

Challenge 693, page 412: At the highest possible mass concentration, entropy is naturally the highest possible.

Challenge 694, page 412: The units do not match.

Challenge 695, page 412: In the case of water, a few turns mixes the ink, and turning backwards increases the mixing. In the case of glycerine, a few turns *seems* to mix the ink, and turning backwards undoes the mixing.

Challenge 696, page 413: Put them in clothes.

Challenge 700, page 413: Negative temperatures are a conceptual crutch definable only for systems with a few discrete states; they are not real temperatures, because they do not describe equilibrium states, and indeed never apply to systems with a continuum of states.

Challenge 701, page 415: This is also true for the shape of human bodies, the brain control of human motion, the growth of flowers, the waves of the sea, the formation of clouds, the processes leading to volcano eruptions, etc.

Page 319 **Challenge 704**, page 422: See the puzzle about the motion of ski moguls.

Challenge 709, page 425: First, there are many more butterflies than tornadoes. Second, tornadoes do not rely on small initial disturbances for their appearance. Third, the belief in the butterfly 'effect' completely neglects an aspect of nature that is essential for self-organization: friction and dissipation. The butterfly 'effect', assumed that it existed, would require that dissipation in the air should have completely unrealistic properties. This is not the case in the atmosphere. But most important of all, there is no experimental basis for the 'effect': it has never been observed. Thus it does not exist.

Challenge 719, page 436: No. Nature does not allow more than about 20 digits of precision, as we will discover later in our walk. That is not sufficient for a standard book. The question whether such a number can be part of its own book thus disappears.

Challenge 720, page 436: All three statements are hogwash. A drag coefficient implies that the cross area of the car is known to the same precision. This is actually extremely difficult to measure and to keep constant. In fact, the value 0.375 for the Ford Escort was a cheat, as many other measurements showed. The fuel consumption is even more ridiculous, as it implies that fuel volumes and distances can be measured to that same precision. Opinion polls are taken by phoning at most 2000 people; due to the difficulties in selecting the right representative sample, that gives a precision of at most 3 % for typical countries.

Challenge 722, page 437: Space-time is defined using matter; matter is defined using space-time.

Challenge 723, page 437: Fact is that physics has been based on a circular definition for hundreds of years. Thus it is possible to build even an exact science on sand. Nevertheless, the elimination of the circularity is an important aim.

Challenge 724, page 438: Every measurement is a comparison with a standard; every comparison requires light or some other electromagnetic field. This is also the case for time measurements.

Challenge 725, page 438: Every mass measurement is a comparison with a standard; every comparison requires light or some other electromagnetic field.

Challenge 726, page 438: Angle measurements have the same properties as length or time measurements.

Challenge 728, page 454: Mass is a measure of the amount of energy. The 'square of mass' makes no sense.

Challenge 731, page 456: About $10\,\mu$g.

Challenge 732, page 457: Probably the quantity with the biggest variation is mass, where a prefix for $1\,eV/c^2$ would be useful, as would be one for the total mass in the universe, which is about 10^{90} times larger.

Challenge 733, page 458: The formula with $n - 1$ is a better fit. Why?

Challenge 736, page 459: No! They are much too precise to make sense. They are only given as an illustration for the behaviour of the Gaussian distribution. Real measurement distributions are not Gaussian to the precision implied in these numbers.

Challenge 737, page 459: About 0.3 m/s. It is *not* 0.33 m/s, it is *not* 0.333 m/s and it is *not* any longer strings of threes!

Challenge 739, page 465: The slowdown goes *quadratically* with time, because every new slowdown adds to the old one!

Challenge 740, page 466: No, only properties of parts of the universe are listed. The universe itself has no properties, as shown in the last volume.   Vol. VI, page 111

Challenge 741, page 528: For example, speed inside materials is slowed, but between atoms, light still travels with vacuum speed.

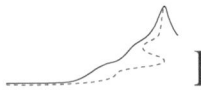

BIBLIOGRAPHY

❝ Aiunt enim multum legendum esse, non multa. ❞
Plinius, *Epistulae*.*

1   For a history of science in antiquity, see LUCIO RUSSO, *La rivoluzione dimenticata*, Feltrinelli, 1996, also available in several other languages. Cited on page 15.

2   If you want to catch up secondary school physics, the clearest and shortest introduction world-wide is a free school text, available in English and several other languages, written by a researcher who has dedicated all his life to the teaching of physics in secondary school, together with his university team: FRIEDRICH HERRMANN, *The Karlsruhe Physics Course*, free to download in English, Spanish, Russian, Italian and Chinese at www.physikdidaktik.uni-karlsruhe.de/index_en.html. It is one of the few secondary school texts that captivates and surprises even professional physicists. (The 2013 paper on this book by C. STRUNK & K. RINCKE, *Zum Gutachten der Deutschen Physikalischen Gesellschaft über den Karlsruher Physikkurs*, available on the internet, makes many interesting points and is enlightening for every physicist.) This can be said even more of the wonderfully daring companion text FRIEDRICH HERRMANN & GEORG JOB, *Historical Burdens on Physics*, whose content is also freely available on the Karlsruhe site, in English and in several other languages.

A beautiful book explaining physics and its many applications in nature and technology vividly and thoroughly is PAUL G. HEWITT, JOHN SUCHOCKI & LESLIE A. HEWITT, *Conceptual Physical Science*, Bejamin/Cummings, 1999.

A great introduction is KLAUS DRANSFELD, PAUL KIENLE & GEORG KALVIUS, *Physik 1: Mechanik und Wärme*, Oldenburg, 2005.

A book series famous for its passion for curiosity is RICHARD P. FEYNMAN, ROBERT B. LEIGHTON & MATTHEW SANDS, *The Feynman Lectures on Physics*, Addison Wesley, 1977. The volumes can now be read online for free at www.feynmanlectures.info.

A lot can be learned about motion from quiz books. One of the best is the well-structured collection of beautiful problems that require no mathematics, written by JEAN-MARC LÉVY-LEBLOND, *La physique en questions – mécanique*, Vuibert, 1998.

Another excellent quiz collection is YAKOV PERELMAN, *Oh, la physique*, Dunod, 2000, a translation from the Russian original.

A good problem book is W. G. REES, *Physics by Example: 200 Problems and Solutions*, Cambridge University Press, 1994.

---

* 'Read much, but not anything.' Ep. 7, 9, 15. Gaius Plinius Secundus (b. 23/4 Novum Comum, d. 79 Vesuvius eruption), Roman writer, especially famous for his large, mainly scientific work *Historia naturalis*, which has been translated and read for almost 2000 years.

A good history of physical ideas is given in the excellent text by DAVID PARK, *The How and the Why*, Princeton University Press, 1988.

An excellent introduction into physics is ROBERT POHL, *Pohl's Einführung in die Physik*, Klaus Lüders & Robert O. Pohl editors, Springer, 2004, in two volumes with CDs. It is a new edition of a book that is over 70 years old; but the didactic quality, in particular of the experimental side of physics, is unsurpassed.

Another excellent Russian physics problem book, the so-called *Saraeva*, seems to exist only as Spanish translation: B.B. BÚJOVTSEV, V.D. KRÍVCHENKOV, G.YA. MIÁK-ISHEV & I.M. SARÁEVA *Problemas seleccionados de física elemental*, Mir, 1979.

Another good physics problem book is GIOVANNI TONZIG, *Cento errori di fisica pronti per l'uso*, Sansoni, third edition, 2006. See also his www.giovannitonzig.it website. Cited on pages 15, 120, 219, 324, and 533.

3 An overview of motion illusions can be found on the excellent website www.michaelbach.de/ot. The complex motion illusion figure is found on www.michaelbach.de/ot/mot_rotsnake/index.html; it is a slight variation of the original by Kitaoka Akiyoshi at www.ritsumei.ac.jp/~akitaoka/rotsnake.gif, published as A. KITAOKA & H. ASHIDA, *Phenomenal characteristics of the peripheral drift illusion*, Vision 15, pp. 261–262, 2003. A common scam is to claim that the illusion is due to or depends on stress. Cited on page 16.

4 These and other fantastic illusions are also found in AKIYOSHI KITAOKA, *Trick Eyes*, Barnes & Noble, 2005. Cited on page 16.

5 A well-known principle in the social sciences states that, given a question, for every possible answer, however weird it may seem, there is somebody – and often a whole group – who holds it as his opinion. One just has to go through literature (or the internet) to confirm this.

About group behaviour in general, see R. AXELROD, *The Evolution of Cooperation*, Harper Collins, 1984. The propagation and acceptance of ideas, such as those of physics, are also an example of human cooperation, with all its potential dangers and weaknesses. Cited on page 16.

6 All the known texts by Parmenides and Heraclitus can be found in JEAN-PAUL DUMONT, *Les écoles présocratiques*, Folio-Gallimard, 1988. Views about the non-existence of motion have also been put forward by much more modern and much more contemptible authors, such as in 1710 by Berkeley. Cited on page 17.

7 An example of people worried by Zeno is given by WILLIAM MCLAUGHLIN, *Resolving Zeno's paradoxes*, Scientific American pp. 66–71, November 1994. The actual argument was not about a hand slapping a face, but about an arrow hitting the target. See also Ref. 65. Cited on page 17.

8 The full text of *La Beauté* and the other poems from *Les fleurs du mal*, one of the finest books of poetry ever written, can be found at the hypermedia.univ-paris8.fr/bibliotheque/Baudelaire/Spleen.html website. Cited on page 18.

9 A famous collection of interesting examples of motion in everyday life is the excellent book by JEARL WALKER, *The Flying Circus of Physics*, Wiley, 1975. Its website is at www.flyingcircusofphysics.com. Another beautiful book is CHRISTIAN UCKE & H. JOACHIM SCHLICHTING, *Spiel, Physik und Spaß – Physik zum Mitdenken und Mitmachen*, Wiley-VCH, 2011. For more interesting physical effects in everyday life, see ERWEIN FLACHSEL, *Hundertfünfzig Physikrätsel*, Ernst Klett Verlag, 1985. The book also covers several clock puzzles, in puzzle numbers 126 to 128. Cited on page 19.

10 A concise and informative introduction into the history of classical physics is given in the first chapter of the book by FLOYD KARKER RICHTMYER, EARLE HESSE KENNARD

Vol. III, page 324

& JOHN N. COOPER, *Introduction to Modern Physics*, McGraw–Hill, 1969. Cited on page 19.

11 An introduction into perception research is E. BRUCE GOLDSTEIN, *Perception*, Books/Cole, 5th edition, 1998. Cited on pages 21 and 26.

12 A good overview over the arguments used to prove the existence of god from motion is given by MICHAEL BUCKLEY, *Motion and Motion's God*, Princeton University Press, 1971. The intensity of the battles waged around these failed attempts is one of the tragicomic chapters of history. Cited on page 21.

13 THOMAS AQUINAS, *Summa Theologiae* or *Summa Theologica*, 1265–1273, online in Latin at www.newadvent.org/summa, in English on several other servers. Cited on page 21.

14 For an exploration of 'inner' motions, see the beautiful text by RICHARD SCHWARTZ, *Internal Family Systems Therapy*, The Guilford Press, 1995. Cited on page 21.

15 For an authoritative description of proper motion development in babies and about how it leads to a healthy character see EMMI PIKLER, *Laßt mir Zeit - Die selbstständige Bewegungsentwicklung des Kindes bis zum freien Gehen*, Pflaum Verlag, 2001, and her other books. See also the website www.pikler.org. Cited on page 21.

16 See e.g. the fascinating text by DAVID G. CHANDLER, *The Campaigns of Napoleon – The Mind and Method of History's Greatest Soldier*, Macmillan, 1966. Cited on page 21.

17 RICHARD MARCUS, *American Roulette*, St Martin's Press, 2003, a thriller and a true story. Cited on page 21.

18 A good and funny book on behaviour change is the well-known text RICHARD BANDLER, *Using Your Brain for a Change*, Real People Press, 1985. See also RICHARD BANDLER & JOHN GRINDER, *Frogs into princes – Neuro Linguistic Programming*, Eden Grove Editions, 1990. Cited on pages 21 and 32.

19 A beautiful book about the mechanisms of human growth from the original cell to full size is LEWIS WOLPERT, *The Triumph of the Embryo*, Oxford University Press, 1991. Cited on page 21.

20 On the topic of grace and poise, see e.g. the numerous books on the Alexander technique, such as M. GELB, *Body Learning - An Introduction to the Alexander Technique*, Aurum Press, 1981, and RICHARD BRENNAN, *Introduction to the Alexander Technique*, Little Brown and Company, 1996. Among others, the idea of the Alexander technique is to return to the situation that the muscle groups for sustainment and those for motion are used only for their respective function, and not vice versa. Any unnecessary muscle tension, such as neck stiffness, is a waste of energy due to the use of sustainment muscles for movement and of motion muscles for sustainment. The technique teaches the way to return to the natural use of muscles.

Motion of animals was discussed extensively already in the seventeenth century by G. BORELLI, *De motu animalium*, 1680. An example of a more modern approach is J. J. COLLINS & I. STEWART, *Hexapodal gaits and coupled nonlinear oscillator models*, Biological Cybernetics 68, pp. 287–298, 1993. See also I. STEWART & M. GOLUBITSKY, *Fearful Symmetry*, Blackwell, 1992. Cited on pages 23 and 122.

21 The results on the development of children mentioned here and in the following have been drawn mainly from the studies initiated by Jean Piaget; for more details on child development, see later on. At www.piaget.org you can find the website maintained by the Jean Piaget Society. Cited on pages 24, 40, and 42.

22 The reptilian brain (eat? flee? ignore?), also called the R-complex, includes the brain stem, the cerebellum, the basal ganglia and the thalamus; the old mammalian (emotions) brain,

also called the limbic system, contains the amygdala, the hypothalamus and the hippocampus; the human (and primate) (rational) brain, called the neocortex, consists of the famous grey matter. For images of the brain, see the atlas by JOHN NOLTE, *The Human Brain: An Introduction to its Functional Anatomy*, Mosby, fourth edition, 1999. Cited on page 25.

23   The lower left corner film can be reproduced on a computer after typing the following lines in the Mathematica software package: Cited on page 26.

```
« Graphics'Animation'
Nxpixels=72; Nypixels=54; Nframes=Nxpixels 4/3;
Nxwind=Round[Nxpixels/4]; Nywind=Round[Nypixels/3];
front=Table[Round[Random[]],{y,1,Nypixels},{x,1,Nxpixels}];
back =Table[Round[Random[]],{y,1,Nypixels},{x,1,Nxpixels}];
frame=Table[front,{nf,1,Nframes}];
Do[ If[ x>n-Nxwind && x<n && y>Nywind && y<2Nywind,
    frame[[n,y,x]]=back[[y,x-n]] ],
        {x,1,Nxpixels}, {y,1,Nypixels}, {n,1,Nframes}];
film=Table[ListDensityPlot[frame[[nf]], Mesh-> False,
    Frame-> False, AspectRatio-> N[Nypixels/Nxpixels],
    DisplayFunction-> Identity],    {nf,1,Nframes}]
ShowAnimation[film]
```

But our motion detection system is much more powerful than the example shown in the lower left corners. The following, different film makes the point.

```
« Graphics'Animation'
Nxpixels=72; Nypixels=54; Nframes=Nxpixels 4/3;
Nxwind=Round[Nxpixels/4]; Nywind=Round[Nypixels/3];
front=Table[Round[Random[]],{y,1,Nypixels},{x,1,Nxpixels}];
back =Table[Round[Random[]],{y,1,Nypixels},{x,1,Nxpixels}];
frame=Table[front,{nf,1,Nframes}];
Do[ If[ x>n-Nxwind && x<n && y>Nywind && y<2Nywind,
    frame[[n,y,x]]=back[[y,x]] ],
        {x,1,Nxpixels}, {y,1,Nypixels}, {n,1,Nframes}];
film=Table[ListDensityPlot[frame[[nf]], Mesh-> False,
    Frame-> False, AspectRatio-> N[Nypixels/Nxpixels],
    DisplayFunction-> Identity],    {nf,1,Nframes}]
ShowAnimation[film]
```

Similar experiments, e.g. using randomly changing random patterns, show that the eye perceives motion even in cases where all Fourier components of the image are practically zero; such image motion is called *drift-balanced* or *non-Fourier* motion. Several examples are presented in J. ZANKER, *Modelling human motion perception I: Classical stimuli*, Naturwissenschaften 81, pp. 156–163, 1994, and J. ZANKER, *Modelling human motion perception II: Beyond Fourier motion stimuli*, Naturwissenschaften 81, pp. 200–209, 1994. Modern research has helped to find the corresponding neuronal structures, as shown in S. A. BACCUS, B. P. OLVECZKY, M. MANU & M. MEISTER, *A retinal circuit that computes object motion*, Journal of Neuroscience 28, pp. 6807–6817, 2008.

24   All fragments from Heraclitus are from JOHN MANSLEY ROBINSON, *An Introduction to Early Greek Philosophy*, Houghton Muffin 1968, chapter 5. Cited on page 27.

25   On the block and tackle, see the explanations by Donald Simanek at http://www.lhup.edu/~dsimanek/TTT-fool/fool.htm. Cited on page 31.

26 An overview over these pretty puzzles is found in E. D. DEMAINE, M. L. DEMAINE, Y. N. MINSKI, J. S. B. MITCHELL, R. L. RIVEST & M. PATRASCU, *Picture-hanging puzzles*, preprint at arxiv.org/abs/1203.3602. Cited on page 33.

27 An introduction to Newton the alchemist are the books by BETTY JO TEETER DOBBS, *The Foundations of Newton's Alchemy*, Cambridge University Press, 1983, and *The Janus Face of Genius*, Cambridge University Press, 1992. Newton is found to be a sort of highly intellectual magician, desperately looking for examples of processes where gods interact with the material world. An intense but tragic tale. A good overview is provided by R. G. KEESING, *Essay Review: Newton's Alchemy*, Contemporary Physics 36, pp. 117–119, 1995.

   Newton's infantile theology, typical for god seekers who grew up without a father, can be found in the many books summarizing the letter exchanges between Clarke, his secretary, and Leibniz, Newton's rival for fame. Cited on page 34.

28 An introduction to the story of classical mechanics, which also destroys a few of the myths surrounding it – such as the idea that Newton could solve differential equations or that he introduced the expression $F = ma$ – is given by CLIFFORD A. TRUESDELL, *Essays in the History of Mechanics*, Springer, 1968. Cited on pages 34, 178, and 228.

29 The slowness of the effective speed of light inside the Sun is due to the frequent scattering of photons by solar matter. The best estimate of its value is by R. MITALAS & K. R. SILLS, *On the photon diffusion time scale for the Sun*, The Astrophysical Journal 401, pp. 759–760, 1992. They give an average speed of 0.97 cm/s over the whole Sun and a value about 10 times smaller at its centre. Cited on page 36.

30 C. LIU, Z. DUTTON, C. H. BEHROOZI & L. VESTERGAARD HAU, *Observation of coherent optical information storage in an atomic medium using halted light pulses*, Nature 409, pp. 490–493, 2001. There is also a comment on the paper by E. A. CORNELL, *Stopping light in its track*, 409, pp. 461–462, 2001. However, despite the claim, the light pulses of course have *not* been halted. Can you give at least two reasons without even reading the paper, and maybe a third after reading it?

   The work was an improvement on the previous experiment where a group velocity of light of 17 m/s had been achieved, in an ultracold gas of sodium atoms, at nanokelvin temperatures. This was reported by L. VESTERGAARD HAU, S. E. HARRIS, Z. DUTTON & C. H. BEHROOZI, *Light speed reduction to 17 meters per second in an ultracold atomic gas*, Nature 397, pp. 594–598, 1999. Cited on page 36.

31 RAINER FLINDT, *Biologie in Zahlen – Eine Datensammlung in Tabellen mit über 10.000 Einzelwerten*, Spektrum Akademischer Verlag, 2000. Cited on page 36.

32 Two jets with that speed have been observed by I. F. MIRABEL & L. F. RODRÍGUEZ, *A superluminal source in the Galaxy*, Nature 371, pp. 46–48, 1994, as well as the comments on p. 18. Cited on page 36.

33 A beautiful introduction to the slowest motions in nature, the changes in landscapes, is DETLEV BUSCHE, JÜRGEN KEMPF & INGRID STENGEL, *Landschaftsformen der Erde – Bildatlas der Geomorphologie*, Primus Verlag, 2005. Cited on page 37.

34 To build your own sundial, see the pretty and short ARNOLD ZENKERT, *Faszination Sonnenuhr*, VEB Verlag Technik, 1984. See also the excellent and complete introduction into this somewhat strange world at the www.sundials.co.uk website. Cited on page 42.

35 An introduction to the sense of time as a result of clocks in the brain is found in R. B. IVRY & R. SPENCER, *The neural representation of time*, Current Opinion in Neurobiology 14, pp. 225–232, 2004. The chemical clocks in our body are described in JOHN D. PALMER,

Challenge 741 s

The Living Clock, Oxford University Press, 2002, or in A. AHLGREN & F. HALBERG, *Cycles of Nature: An Introduction to Biological Rhythms*, National Science Teachers Association, 1990. See also the www.msi.umn.edu/~halberg/introd website. Cited on page 43.

36   This has been shown among others by the work of Anna Wierzbicka that is discussed in more detail in one of the subsequent volumes. The passionate best seller by the Chomskian author STEVEN PINKER, *The Language Instinct – How the Mind Creates Language*, Harper Perennial, 1994, also discusses issues related to this matter, refuting amongst others on page 63 the often repeated false statement that the *Hopi* language is an exception. Cited on page 43. Vol. III, page 279

37   For more information, see the excellent and freely downloadable books on biological clocks by Wolfgang Engelmann on the website www.uni-tuebingen.de/plantphys/bioclox. Cited on page 44.

38   B. GÜNTHER & E. MORGADO, *Allometric scaling of biological rhythms in mammals*, Biological Research 38, pp. 207–212, 2005. Cited on page 44.

39   Aristotle rejects the idea of the flow of time in chapter IV of his *Physics*. See the full text on the classics.mit.edu/Aristotle/physics.4.iv.html website. Cited on page 48.

40   Perhaps the most informative of the books about the 'arrow of time' is HANS DIETER ZEH, *The Physical Basis of the Direction of Time*, Springer Verlag, 4th edition, 2001. It is still the best book on the topic. Most other texts exist – have a look on the internet – but lack clarity of ideas.

   A typical conference proceeding is J. J. HALLIWELL, J. PÉREZ-MERCADER & WOJCIECH H. ZUREK, *Physical Origins of Time Asymmetry*, Cambridge University Press, 1994. Cited on page 49.

41   On the issue of absolute and relative motion there are many books about few issues. Examples are JULIAN BARBOUR, *Absolute or Relative Motion? Vol. 1: A Study from the Machian Point of View of the Discovery and the Structure of Spacetime Theories*, Cambridge University Press, 1989, JULIAN BARBOUR, *Absolute or Relative Motion? Vol. 2: The Deep Structure of General Relativity*, Oxford University Press, 2005, or JOHN EARMAN, *World Enough and Spacetime: Absolute vs Relational Theories of Spacetime*, MIT Press, 1989. A speculative solution on the alternative between absolute and relative motion is presented in volume VI. Cited on page 52. Vol. VI, page 65

42   Coastlines and other fractals are beautifully presented in HEINZ-OTTO PEITGEN, HARTMUT JÜRGENS & DIETMAR SAUPE, *Fractals for the Classroom*, Springer Verlag, 1992, pp. 232–245. It is also available in several other languages. Cited on page 54.

43   R. DOUGHERTY & M. FOREMAN, *Banach–Tarski decompositions using sets with the property of Baire*, Journal of the American Mathematical Society 7, pp. 75–124, 1994. See also ALAN L.T. PATERSON, *Amenability*, American Mathematical Society, 1998, and ROBERT M. FRENCH, *The Banach–Tarski theorem*, The Mathematical Intelligencer 10, pp. 21–28, 1998. Finally, there are the books by BERNARD R. GELBAUM & JOHN M. H. OLMSTED, *counter-examples in Analysis*, Holden–Day, 1964, and their *Theorems and counter-examples in Mathematics*, Springer, 1993. Cited on page 57.

44   The beautiful but not easy text is STAN WAGON, *The Banach Tarski Paradox*, Cambridge University Press, 1993. Cited on pages 58 and 481.

45   About the shapes of salt water bacteria, see the corresponding section in the interesting book by BERNARD DIXON, *Power Unseen – How Microbes Rule the World*, W.H. Freeman, 1994. The book has about 80 sections, in which as many microorganisms are vividly presented. Cited on page 59.

46 OLAF MEDENBACH & HARRY WILK, *Zauberwelt der Mineralien*, Sigloch Edition, 1977. It combines beautiful photographs with an introduction into the science of crystals, minerals and stones. About the largest crystals, see P. C. RICKWOOD, *The largest crystals*, 66, pp. 885–908, 1981, also available on www.minsocam.org/MSA/collectors_corner/arc/large_crystals.htm. For an impressive example, the Naica cave in Mexico, see www.naica.com.mx/ingles/index.htm Cited on page 59.

47 See the websites www.weltbildfrage.de/3frame.htm and www.lhup.edu/~dsimanek/hollow/morrow.htm. Cited on page 60.

48 The smallest distances are probed in particle accelerators; the distance can be determined from the energy of the particle beam. In 1996, the value of 10^{-19} m (for the upper limit of the size of quarks) was taken from the experiments described in F. ABE & al., *Measurement of dijet angular distributions by the collider detector at Fermilab*, Physical Review Letters 77, pp. 5336–5341, 1996. Cited on page 66.

49 More on the Moon illusion can be found at the website science.nasa.gov/science-news/science-at-nasa/2008/16jun_moonillusion/. All the works of Ptolemy are found online at www.ptolemaeus.badw.de. Cited on page 69.

50 These puzzles are taken from the puzzle collection at www.mathematische-basteleien.de. Cited on page 70.

51 ALEXANDER K. DEWDNEY, *The Planiverse – Computer Contact with a Two-dimensional World*, Poseidon Books/Simon & Schuster, 1984. See also EDWIN A. ABBOTT, *Flatland: A romance of many dimensions*, 1884. Several other fiction authors had explored the option of a two-dimensional universe before, always answering, incorrectly, in the affirmative. Cited on page 71.

52 J. BOHR & K. OLSEN, *The ancient art of laying rope*, preprint at arxiv.org/abs/1004.0814 Cited on page 71.

53 For an overview and references see www.pbrc.hawaii.edu/~petra/animal_olympians.html. Cited on page 72.

54 P. PIERANSKI, S. PRZYBYL & A. STASIAK, *Tight open knots*, European Physical Journal E 6, pp. 123–128, 2001, preprint at arxiv.org/abs/physics/0103016. Cited on page 73.

55 On the world of fireworks, see the frequently asked questions list of the usenet group rec.pyrotechnics, or search the web. A simple introduction is the article by J. A. CONKLING, *Pyrotechnics*, Scientific American pp. 66–73, July 1990. Cited on page 75.

56 There is a whole story behind the variations of g. It can be discovered in CHUJI TSUBOI, *Gravity*, Allen & Unwin, 1979, or in WOLFGANG TORGE, *Gravimetry*, de Gruyter, 1989, or in MILAN BURŠA & KAREL PĚČ, *The Gravity Field and the Dynamics of the Earth*, Springer, 1993. The variation of the height of the soil by up to 0.3 m due to the Moon is one of the interesting effects found by these investigations. Cited on pages 76 and 197.

57 STILLMAN DRAKE, *Galileo: A Very Short Introduction*, Oxford University Press, 2001. Cited on page 76.

58 ANDREA FROVA, *La fisica sotto il naso – 44 pezzi facili*, Biblioteca Universale Rizzoli, Milano, 2001. Cited on page 77.

59 On the other hands, other sciences enjoy studying usual paths in all detail. See, for example, HEINI HEDIGER, editor, *Die Straßen der Tiere*, Vieweg & Sohn, 1967. Cited on page 77.

60 H. K. ERIKSEN, J. R. KRISTIANSEN, Ø. LANGANGEN & I. K. WEHUS, *How fast could Usain Bolt have run? A dynamical study*, American Journal of Physics 77, pp. 224–228, 2009. See also the references at en.wikipedia.org/wiki/Footspeed. Cited on page 78.

61 This was discussed in the *Frankfurter Allgemeine Zeitung*, 2nd of August, 1997, at the time of the world athletics championship. The values are for the fastest part of the race of a 100 m sprinter; the exact values cited were called the running speed world records in 1997, and were given as 12.048 m/s = 43.372 km/h by Ben Johnson for men, and 10.99 m/s = 39.56 km/h for women. Cited on page 78.

62 Long jump data and literature can be found in three articles all entitled *Is a good long jumper a good high jumper?*, in the American Journal of Physics 69, pp. 104–105, 2001. In particular, world class long jumpers run at 9.35 ± 0.15 m/s, with vertical take-off speeds of 3.35 ± 0.15 m/s, giving take-off angles of about (only) 20°. A new technique for achieving higher take-off angles would allow the world long jump record to increase dramatically. Cited on page 78.

63 The study of shooting faeces (i.e., shit) and its mechanisms is a part of modern biology. The reason that caterpillars do this was determined by M. WEISS, *Good housekeeping: why do shelter-dwelling caterpillars fling their frass?*, Ecology Letters 6, pp. 361–370, 2003, who also gives the present record of 1.5 m for the 24 mg pellets of *Epargyreus clarus*. The picture of the flying frass is from S. CAVENEY, H. MCLEAN & D. SURRY, *Faecal firing in a skipper caterpillar is pressure-driven*, The Journal of Experimental Biology 201, pp. 121–133, 1998. Cited on page 79.

64 H. C. BENNET-CLARK, *Scale effects in jumping animals*, pp. 185–201, in T. J. PEDLEY, editor, *Scale Effects in Animal Locomotion*, Academic Press, 1977. Cited on page 80.

65 The arguments of Zeno can be found in ARISTOTLE, *Physics*, VI, 9. It can be found translated in almost any language. The classics.mit.edu/Aristotle/physics.6.vi.html website provides an online version in English. Cited on pages 83 and 525.

66 See, for exaple, K. V. KUMAR & W. T. NORFLEET, *Issues of human acceleration tolerance after long-duration space flights*, NASA Technical Memorandum 104753, pp. 1–55, 1992, available at ntrs.nasa.gov. Cited on page 85.

67 Etymology can be a fascinating topic, e.g. when research discovers the origin of the German word 'Weib' ('woman', related to English 'wife'). It was discovered, via a few texts in Tocharian – an extinct Indo-European language from a region inside modern China – to mean originally 'shame'. It was used for the female genital region in an expression meaning 'place of shame'. With time, this expression became to mean 'woman' in general, while being shortened to the second term only. This connection was discovered by the linguist Klaus T. Schmidt; it explains in particular why the word is not feminine but neutral, i.e., why it uses the article 'das' instead of 'die'. Julia Simon, private communication.

Etymology can also be simple and plain fun, for example when one discovers in the *Oxford English Dictionary* that 'testimony' and 'testicle' have the same origin; indeed in Latin the same word 'testis' was used for both concepts. Cited on pages 86 and 101.

68 An overview of the latest developments is given by J. T. ARMSTRONG, D. J. HUNTER, K. J. JOHNSTON & D. MOZURKEWICH, *Stellar optical interferometry in the 1990s*, Physics Today pp. 42–49, May 1995. More than 100 stellar diameters were known already in 1995. Several dedicated powerful instruments are being planned. Cited on page 87.

69 A good biology textbook on growth is ARTHUR F. HOPPER & NATHAN H. HART, *Foundations of Animal Deveopment*, Oxford University Press, 2006. Cited on page 89.

70 This is discussed for example in C. L. STONG, *The amateur scientist – how to supply electric power to something which is turning*, Scientific American pp. 120–125, December 1975. It also discusses how to make a still picture of something rotating simply by using a few prisms, the so-called *Dove prisms*. Other examples of attaching something to a rotating body are given by E. RIEFLIN, *Some mechanisms related to Dirac's strings*, American Journal of Physics 47, pp. 379–381, 1979. Cited on page 89.

71   JAMES A. YOUNG, *Tumbleweed*, Scientific American 264, pp. 82–87, March 1991. The tumbleweed is in fact quite rare, except in Hollywood westerns, where all directors feel obliged to give it a special appearance. Cited on page 90.

72   The classic book on the topic is JAMES GRAY, *Animal Locomotion*, Weidenfeld & Nicolson, 1968. Cited on page 90.

73   About *N. decemspinosa*, see R. L. CALDWELL, *A unique form of locomotion in a stomatopod – backward somersaulting*, Nature 282, pp. 71–73, 1979, and R. FULL, K. EARLS, M. WONG & R. CALDWELL, *Locomotion like a wheel?*, Nature 365, p. 495, 1993. About rolling caterpillars, see J. BRACKENBURY, *Caterpillar kinematics*, Nature 330, p. 453, 1997, and J. BRACKENBURY, *Fast locomotion in caterpillars*, Journal of Insect Physiology 45, pp. 525–533, 1999. More images around legs can be found on rjf9.biol.berkeley.edu/twiki/bin/view/PolyPEDAL/LabPhotographs. Cited on page 90.

74   The locomotion of the spiders of the species *Cebrennus villosus* has been described by Ingo Rechenberg from Berlin. See the video at www.youtube.com/watch?v=Aayb_h31RyQ. Cited on page 90.

75   The first experiments to prove the rotation of the flagella were by M. SILVERMAN & M. I. SIMON, *Flagellar rotation and the mechanism of bacterial motility*, Nature 249, pp. 73–74, 1974. For some pretty pictures of the molecules involved, see K. NAMBA, *A biological molecular machine: bacterial flagellar motor and filament*, Wear 168, pp. 189–193, 1993, or the website www.nanonet.go.jp/english/mailmag/2004/011a.html. The present record speed of rotation, 1700 rotations per second, is reported by Y. MAGARIYAMA, S. SUGIYAMA, K. MURAMOTO, Y. MAEKAWA, I. KAWAGISHI, Y. IMAE & S. KUDO, *Very fast flagellar rotation*, Nature 371, p. 752, 1994.

More on bacteria can be learned from DAVID DUSENBERY, *Life at a Small Scale*, Scientific American Library, 1996. Cited on page 92.

76   S. CHEN & al., *Structural diversity of bacterial flagellar motors*, EMBO Journal 30, pp. 2972–2981, 2011, also online at emboj.embopress.org/content/30/14/2972. Cited on page 92.

77   M. P. BRENNER, S. HILGENFELDT & D. LOHSE, *Single bubble sonoluminescence*, Reviews of Modern Physics 74, pp. 425–484, 2002. Cited on page 96.

78   K. R. WENINGER, B. P. BARBER & S. J. PUTTERMAN, *Pulsed Mie scattering measurements of the collapse of a sonoluminescing bubble*, Physical Review Letters 78, pp. 1799–1802, 1997. Cited on page 96.

79   On shadows, see the agreeable popular text by ROBERTO CASATI, *Alla scoperta dell'ombra – Da Platone a Galileo la storia di un enigma che ha affascinato le grandi menti dell'umanità*, Oscar Mondadori, 2000, and his website located at www.shadowes.org. Cited on page 98.

80   There is also the beautiful book by PENELOPE FARRANT, *Colour in Nature*, Blandford, 1997. Cited on page 98.

81   The 'laws' of cartoon physics can easily be found using any search engine on the internet. Cited on page 98.

82   For the curious, an overview of the illusions used in the cinema and in television, which lead to some of the strange behaviour of images mentioned above, is given in BERNARD WILKIE, *The Technique of Special Effects in Television*, Focal Press, 1993, and his other books, or in the *Cinefex* magazine. On digital cinema techniques, see PETER C. SLANSKY, editor, *Digitaler film – digitales Kino*, UVK Verlag, 2004. Cited on page 99.

83 AETIUS, Opinions, I, XXIII, 3. See JEAN-PAUL DUMONT, *Les écoles présocratiques*, Folio Essais, Gallimard, p. 426, 1991. Cited on page 99.

84 GIUSEPPE FUMAGALLI, *Chi l'ha detto?*, Hoepli, 1983. It is from Pappus of Alexandria's opus *Synagoge*, book VIII, 19. Cited on pages 100 and 237.

85 See www.straightdope.com/classics/a5_262.html and the more dubious en.wikipedia.org/wiki/Guillotine. Cited on page 102.

86 See the path-breaking paper by A. DiSESSA, *Momentum flow as an alternative perspective in elementary mechanics*, 48, p. 365, 1980, and A. DiSESSA, *Erratum: "Momentum flow as an alternative perspective in elementary mechanics"* [Am. J. Phys. 48, 365 (1980)], 48, p. 784, 1980. Also the wonderful free textbook by FRIEDRICH HERRMANN, *The Karlsruhe Physics Course*, makes this point extensively; see Ref. 2. Cited on pages 109, 228, 231, and 508.

87 For the role and chemistry of adenosine triphosphate (ATP) in cells and in living beings, see any chemistry book, or search the internet. The uncovering of the mechanisms around ATP has led to Nobel Prizes in Chemistry in 1978 and in 1997. Cited on page 109.

88 A picture of this unique clock can be found in the article by A. GARRETT, *Perpetual motion – a delicious delirium*, Physics World pp. 23–26, December 1990. Cited on page 110.

89 ESGER BRUNNER, *Het ongelijk van Newton – het kleibakexperiment van 's Gravesande nagespeld*, Nederland tijdschrift voor natuurkunde pp. 95–96, Maart 2012. The paper contains photographs of the mud imprints. Cited on page 111.

90 A Shell study estimated the world's total energy consumption in 2000 to be 500 EJ. The US Department of Energy estimated it to be around 416 EJ. We took the lower value here. A discussion and a breakdown into electricity usage (14 EJ) and other energy forms, with variations per country, can be found in S. BENKA, *The energy challenge*, Physics Today 55, pp. 38–39, April 2002, and in E. J. MONITZ & M. A. KENDERDINE, *Meeting energy challenges: technology and policy*, Physics Today 55, pp. 40–46, April 2002. Cited on pages 113 and 114.

91 L. M. MILLER, F. GANS & A. KLEIDON, *Estimating maximum global land surface wind power extractability and associated climatic consequences*, Earth System Dynamics 2, pp. 1–12, 2011. Cited on page 114.

92 For an overview, see the paper by J. F. MULLIGAN & H. G. HERTZ, *An unpublished lecture by Heinrich Hertz: 'On the energy balance of the Earth'*, American Journal of Physics 65, pp. 36–45, 1997. Cited on page 114.

93 For a beautiful photograph of this feline feat, see the cover of the journal and the article of J. DARIUS, *A tale of a falling cat*, Nature 308, p. 109, 1984. Cited on page 121.

94 NATTHI L. SHARMA, *A new observation about rolling motion*, European Journal of Physics 17, pp. 353–356, 1996. Cited on page 121.

95 C. SINGH, *When physical intuition fails*, American Journal of Physics 70, pp. 1103–1109, 2002. Cited on page 121.

96 There is a vast literature on walking. Among the books on the topic, two well-known introductions are ROBERT MCNEILL ALEXANDER, *Exploring Biomechanics: Animals in Motion*, Scientific American Library, 1992, and STEVEN VOGEL, *Comparative Biomechanics - Life's Physical World*, Princeton University Press, 2003. Cited on page 122.

97 SERGE GRACOVETSKY, *The Spinal Engine*, Springer Verlag, 1990. It is now also known that human gait is chaotic. This is explained by M. PERC, *The dynamics of human gait*, European Journal of Physics 26, pp. 525–534, 2005. On the physics of walking and running, see also the respective chapters in the delightful book by WERNER GRUBER, *Unglaublich einfach, einfach unglaublich: Physik für jeden Tag*, Heyne, 2006. Cited on page 123.

98 M. LLOBERA & T. J. SLUCKIN, *Zigzagging: theoretical insights on climbing strategies*, Journal of Theoretical Biology 249, pp. 206–217, 2007. Cited on page 125.

99 This description of life and death is called the concept of *maximal metabolic scope*. Look up details in your favourite library. A different phrasing is the one by M. YA. AZBEL, *Universal biological scaling and mortality*, Proceedings of the National Academy of Sciences of the USA 91, pp. 12453–12457, 1994. He explains that every atom in an organism consumes, on average, 20 oxygen molecules per life-span. Cited on page 125.

100 DUNCAN MACDOUGALL, *Hypothesis concerning soul substance together with experimental evidence of the existence of such substance*, American Medicine 2, pp. 240–243, April 1907, and DUNCAN MACDOUGALL, *Hypothesis concerning soul substance*, American Medicine 2, pp. 395–397, July 1907. Reading the papers shows that the author has little practice in performing reliable weight and time measurements. Cited on page 126.

101 A good roulette prediction story from the 1970s is told by THOMAS A. BASS, *The Eudaemonic Pie* also published under the title *The Newtonian Casino*, Backinprint, 2000. An overview up to 1998 is given in the paper EDWARD O. THORP, *The invention of the first wearable computer*, *Proceedings of the Second International Symposium on Wearable Computers* (ISWC 1998), 19-20 October 1998, Pittsburgh, Pennsylvania, USA (IEEE Computer Society), pp. 4–8, 1998, downloadable at csdl.computer.org/comp/proceedings/iswc/1998/9074/00/9074toc.htm. Cited on page 127.

102 This and many other physics surprises are described in the beautiful lecture script by JOSEF ZWECK, *Physik im Alltag*, the notes of his lectures held in 1999/2000 at the Universität Regensburg. Cited on pages 128 and 133.

103 The equilibrium of ships, so important in car ferries, is an interesting part of shipbuilding; an introduction was already given by LEONHARD EULER, *Scientia navalis*, 1749. Cited on page 129.

104 THOMAS HEATH, *Aristarchus of Samos – the Ancient Copernicus*, Dover, 1981, reprinted from the original 1913 edition. Aristarchus' treaty is given in Greek and English. Aristarchus was the first proposer of the heliocentric system. Aristarchus had measured the length of the day (in fact, by determining the number of days per year) to the astonishing precision of less than one second. This excellent book also gives an overview of Greek astronomy before Aristarchus, explained in detail for each Greek thinker. Aristarchus' text is also reprinted in ARISTARCHUS, *On the sizes and the distances of the Sun and the Moon*, c. 280 BCE in MICHAEL J. CROWE, *Theories of the World From Antiquity to the Copernican Revolution*, Dover, 1990, especially on pp. 27–29. No citations.

105 T. GERKEMA & L. GOSTIAUX, *A brief history of the Coriolis force*, Europhysics News 43, pp. 14–17, 2012. Cited on page 138.

106 See for example the videos on the Coriolis effect at techtv.mit.edu/videos/3722 and techtv.mit.edu/videos/3714, or search for videos on youtube.com. Cited on page 139.

107 The influence of the Coriolis effect on icebergs was studied most thoroughly by the physicist turned oceanographer Walfrid Ekman (b. 1874 Stockholm, d. 1954 Gostad); the topic was suggested by the great explorer Fridtjof Nansen, who also made the first observations. In his honour, one speaks of the Ekman layer, Ekman transport and Ekman spirals. Any text on oceanography or physical geography will give more details about them. Cited on page 139.

108 An overview of the effects of the Coriolis acceleration $a = -2\omega \times v$ in the rotating frame is given by EDWARD A. DESLOGE, *Classical Mechanics*, Volume 1, John Wiley & Sons, 1982. Even the so-called *Gulf Stream*, the current of warm water flowing from the Caribbean to the North Sea, is influenced by it. Cited on page 139.

109   The original publication is by A. H. Shapiro, *Bath-tub vortex*, Nature 196, pp. 1080–1081, 1962. He also produced two films of the experiment. The experiment has been repeated many times in the northern and in the southern hemisphere, where the water drains clockwise; the first southern hemisphere test was L. M. Trefethen & al., *The bath-tub vortex in the southern hemisphere*, Nature 201, pp. 1084–1085, 1965. A complete literature list is found in the letters to the editor of the American Journal of Physics 62, p. 1063, 1994. Cited on page 140.

110   The tricks are explained by H. Richard Crane, *Short Foucault pendulum: a way to eliminate the precession due to ellipticity*, American Journal of Physics 49, pp. 1004–1006, 1981, and particularly in H. Richard Crane, *Foucault pendulum wall clock*, American Journal of Physics 63, pp. 33–39, 1993. The Foucault pendulum was also the topic of the thesis of Heike Kamerling Onnes, *Nieuwe bewijzen der aswenteling der aarde*, Universiteit Groningen, 1879. Cited on page 140.

111   The reference is J. G. Hagen, *La rotation de la terre : ses preuves mécaniques anciennes et nouvelles*, Sp. Astr. Vaticana Second. App. Rome, 1910. His other experiment is published as J. G. Hagen, *How Atwood's machine shows the rotation of the Earth even quantitatively*, International Congress of Mathematics, Aug. 1912. Cited on page 141.

112   The original papers are A. H. Compton, *A laboratory method of demonstrating the Earth's rotation*, Science 37, pp. 803–806, 1913, A. H. Compton, *Watching the Earth revolve*, Scientific American Supplement no. 2047, pp. 196–197, 1915, and A. H. Compton, *A determination of latitude, azimuth and the length of the day independent of astronomical observations*, Physical Review (second series) 5, pp. 109–117, 1915. Cited on page 141.

113   The G-ring in Wettzell is so precise, with a resolution of less than 10^{-8}, that it has detected the motion of the poles. For details, see K. U. Schreiber, A. Velikoseltsev, M. Rothacher, T. Kluegel, G. E. Stedman & D. L. Wiltshire, *Direct measurement of diurnal polar motion by ring laser gyroscopes*, Journal of Geophysical Research 109 B, p. 06405, 2004, an a review article at T. Klügel, W. Schlüter, U. Schreiber & M. Schneider, *Großringlaser zur kontinuierlichen Beobachtung der Erdrotation*, Zeitschrift für Vermessungswesen 130, pp. 99–108, February 2005. Cited on page 142.

114   R. Anderson, H. R. Bilger & G. E. Stedman, *The Sagnac-effect: a century of Earth-rotated interferometers*, American Journal of Physics 62, pp. 975–985, 1994.

   See also the clear and extensive paper by G. E. Stedman, *Ring laser tests of fundamental physics and geophysics*, Reports on Progress in Physics 60, pp. 615–688, 1997. Cited on page 144.

115   About the length of the day, see the maia.usno.navy.mil website, or the books by K. Lambeck, *The Earth's Variable Rotation: Geophysical Causes and Consequences*, Cambridge University Press, 1980, and by W. H. Munk & G. J. F. MacDonald, *The Rotation of the Earth*, Cambridge University Press, 1960. For a modern ring laser set-up, see www.wettzell.ifag.de. Cited on pages 145 and 200.

116   H. Bucka, *Zwei einfache Vorlesungsversuche zum Nachweis der Erddrehung*, Zeitschrift für Physik 126, pp. 98–105, 1949, and H. Bucka, *Zwei einfache Vorlesungsversuche zum Nachweis der Erddrehung. II. Teil*, Zeitschrift für Physik 128, pp. 104–107, 1950. Cited on page 145.

117   One example of data is by C. P. Sonett, E. P. Kvale, A. Zakharian, M. A. Chan & T. M. Demko, *Late proterozoic and paleozoic tides, retreat of the moon, and rotation of the Earth*, Science 273, pp. 100–104, 5 July 1996. They deduce from tidal sediment analysis that days were only 18 to 19 hours long in the Proterozoic, i.e., 900 million years ago; they assume that the year was 31 million seconds long from then to today. See also C. P. Sonett

& M. A. Chan, *Neoproterozoic Earth-Moon dynamics – rework of the 900 MA Big Cotton-wood canyon tidal laminae*, Geophysical Research Letters 25, pp. 539–542, 1998. Another determination was by G. E. Williams, *Precambrian tidal and glacial clastic deposits: implications for precambrian Earth–Moon dynamics and palaeoclimate*, Sedimentary Geology 120, pp. 55–74, 1998. Using a geological formation called *tidal rhythmites*, he deduced that about 600 million years ago there were 13 months per year and a day had 22 hours. Cited on page 145.

118 For the story of this combination of history and astronomy see Richard Stephenson, *Historical Eclipses and Earth's Rotation*, Cambridge University Press, 1996. Cited on page 146.

119 B. F. Chao, *Earth Rotational Variations excited by geophysical fluids*, IVS 2004 General Meeting proceedings/ pages 38-46. Cited on page 146.

120 On the rotation and history of the Solar System, see S. Brush, *Theories of the origin of the solar system 1956–1985*, Reviews of Modern Physics 62, pp. 43–112, 1990. Cited on page 146.

121 The website hpiers.obspm.fr/eop-pc shows the motion of the Earth's axis over the last ten years. The International Latitude Service founded by Küstner is now part of the International Earth Rotation Service; more information can be found on the www.iers.org website. The latest idea is that two-thirds of the circular component of the polar motion, which in the USA is called 'Chandler wobble' after the person who attributed to himself the discovery by Küstner, is due to fluctuations of the ocean pressure at the bottom of the oceans and one-third is due to pressure changes in the atmosphere of the Earth. This is explained by R. S. Gross, *The excitation of the Chandler wobble*, Geophysical Physics Letters 27, pp. 2329–2332, 2000. Cited on page 147.

122 S. B. Lambert, C. Bizouard & V. Dehant, *Rapid variations in polar motion during the 2005-2006 winter season*, Geophysical Research Letters 33, p. L13303, 2006. Cited on page 147.

123 For more information about Alfred Wegener, see the (simple) text by Klaus Rohrbach, *Alfred Wegener – Erforscher der wandernden Kontinente*, Verlag Freies Geistesleben, 1993; about plate tectonics, see the www.scotese.com website. About earthquakes, see the www.geo.ed.ac.uk/quakexe/quakes and the www.iris.edu/seismon website. See the vulcan.wr.usgs.gov and the www.dartmouth.edu/~volcano websites for information about volcanoes. Cited on page 150.

124 J. Jouzel & al., *Orbital and millennial Antarctic climate variability over the past 800,000 years*, Science 317, pp. 793–796, 2007, takes the data from isotope concentrations in ice cores. In contrast, J. D. Hays, J. Imbrie & N. J. Shackleton, *Variations in the Earth's orbit: pacemaker of the ice ages*, Science 194, pp. 1121–1132, 1976, confirmed the connection with orbital parameters by literally digging in the mud that covers the ocean floor in certain places. Note that the web is full of information on the ice ages. Just look up 'Milankovitch' in a search engine. Cited on pages 153 and 154.

125 R. Humphreys & J. Larsen, *The sun's distance above the galactic plane*, Astronomical Journal 110, pp. 2183–2188, November 1995. Cited on page 153.

126 C. L. Bennet, M. S. Turner & M. White, *The cosmic rosetta stone*, Physics Today 50, pp. 32–38, November 1997. Cited on page 155.

127 The website www.geoffreylandis.com/vacuum.html gives a description of what happened. See also the www.geoffreylandis.com/ebullism.html and imagine.gsfc.nasa.gov/docs/ask_astro/answers/970603.html websites. They all give details on the effects of vacuum on humans. Cited on page 161.

128 R. McN. Alexander, *Leg design and jumping technique for humans, other vertebrates and insects*, Philosophical Transactions of the Royal Society in London B 347, pp. 235–249, 1995. Cited on page 169.

129 J. W. Glasheen & T. A. McMahon, *A hydrodynamic model of locomotion in the basilisk lizard*, Nature 380, pp. 340–342, For pictures, see also New Scientist, p. 18, 30 March 1996, or Scientific American, pp. 48–49, September 1997, or the website by the author at rjf2.biol. berkeley.edu/Full_Lab/FL_Personnel/J_Glasheen/J_Glasheen.html.
     Several shore birds also have the ability to run over water, using the same mechanism. Cited on page 169.

130 A. Fernandez-Nieves & F. J. de las Nieves, *About the propulsion system of a kayak and of Basiliscus basiliscus*, European Journal of Physics 19, pp. 125–129, 1998. Cited on page 170.

131 Y. S. Song, S. H. Suhr & M. Sitti, *Modeling of the supporting legs for designing biomimetic water strider robot*, Proceedings of the IEEE International Conference on Robotics and Automation, Orlando, USA, 2006. S. H. Suhr, Y. S. Song, S. J. Lee & M. Sitti, *Biologically inspired miniature water strider robot*, Proceedings of the Robotics: Science and Systems I, Boston, USA, 2005. See also the website www.me.cmu.edu/faculty1/sitti/nano/ projects/waterstrider. Cited on page 170.

132 J. Iriarte-Díaz, *Differential scaling of locomotor performance in small and large terrestrial mammals*, The Journal of Experimental Biology 205, pp. 2897–2908, 2002. Cited on pages 171 and 561.

133 M. Wittlinger, R. Wehner & H. Wolf, *The ant odometer: stepping on stilts and stumps*, Science 312, pp. 1965–1967, 2006. Cited on page 171.

134 P. G. Weyand, D. B. Sternlight, M. J. Bellizzi & S. Wright, *Faster top running speeds are achieved with greater ground forces not more rapid leg movements*, Journal of Applied Physiology 89, pp. 1991–1999, 2000. Cited on page 171.

135 The material on the shadow discussion is from the book by Robert M. Pryce, *Cook and Peary*, Stackpole Books, 1997. See also the details of Peary's forgeries in Wally Herbert, *The Noose of Laurels*, Doubleday 1989. The sad story of Robert Peary is also told in the centenary number of *National Geographic*, September 1988. Since the National Geographic Society had financed Peary in his attempt and had supported him until the US Congress had declared him the first man at the Pole, the (partial) retraction is noteworthy. (The magazine then changed its mind again later on, to sell more copies, and now again claims that Peary reached the North Pole.) By the way, the photographs of Cook, who claimed to have been at the North Pole even before Peary, have the same problem with the shadow length. Both men have a history of cheating about their 'exploits'. As a result, the first man at the North Pole was probably Roald Amundsen, who arrived there a few years later, and who was also the first man at the South Pole. Cited on page 135.

136 The story is told in M. Nauenberg, *Hooke, orbital motion, and Newton's Principia*, American Journal of Physics 62, 1994, pp. 331–350. Cited on page 178.

137 More details are given by D. Rawlins, in *Doubling your sunsets or how anyone can measure the Earth's size with wristwatch and meter stick*, American Journal of Physics 47, 1979, pp. 126–128. Another simple measurement of the Earth radius, using only a sextant, is given by R. O'Keefe & B. Ghavimi-Alagha, in *The World Trade Center and the distance to the world's center*, American Journal of Physics 60, pp. 183–185, 1992. Cited on page 179.

138 More details on astronomical distance measurements can be found in the beautiful little book by A. van Helden, *Measuring the Universe*, University of Chicago Press, 1985, and in NIGEL HENBEST & HEATHER COOPER, *The Guide to the Galaxy*, Cambridge University Press, 1994. Cited on page 179.

139 A lot of details can be found in M. JAMMER, *Concepts of Mass in Classical and Modern Physics*, reprinted by Dover, 1997, and in *Concepts of Force, a Study in the Foundations of Mechanics*, Harvard University Press, 1957. These eclectic and thoroughly researched texts provide numerous details and explain various philosophical viewpoints, but lack clear statements and conclusions on the accurate description of nature; thus are not of help on fundamental issues.

Jean Buridan (*c.* 1295 to *c.* 1366) criticizes the distinction of sublunar and translunar motion in his book *De Caelo*, one of his numerous works. Cited on page 180.

140 D. TOPPER & D. E. VINCENT, *An analysis of Newton's projectile diagram*, European Journal of Physics 20, pp. 59–66, 1999. Cited on page 180.

141 The absurd story of the metre is told in the historical novel by KEN ALDER, *The Measure of All Things : The Seven-Year Odyssey and Hidden Error that Transformed the World*, The Free Press, 2003. Cited on page 183.

142 H. CAVENDISH, *Experiments to determine the density of the Earth*, Philosophical Transactions of the Royal Society 88, pp. 469–526, 1798. In fact, the first value of the gravitational constant G found in the literature is only from 1873, by Marie-Alfred Cornu and Jean-Baptistin Baille, who used an improved version of Cavendish's method. Cited on page 185.

143 About the measurement of spatial dimensions via gravity – and the failure to find any hint for a number different from three – see the review by E. G. ADELBERGER, B. R. HECKEL & A. E. NELSON, *Tests of the gravitational inverse-square law*, Annual Review of Nuclear and Particle Science 53, pp. 77–121, 2003, also arxiv.org/abs/hep-ph/0307284, or the review by J. A. HEWETT & M. SPIROPULU, *Particle physics probes of extra spacetime dimensions*, Annual Review of Nuclear and Particle Science 52, pp. 397–424, 2002, arxiv.org/abs/hep-ph/0205106. Cited on page 188.

144 There are many books explaining the origin of the precise shape of the Earth, such as the pocket book S. ANDERS, *Weil die Erde rotiert*, Verlag Harri Deutsch, 1985. Cited on page 188.

145 The shape of the Earth is described most precisely with the World Geodetic System. For a presentation, see the en.wikipedia.org/wiki/World_Geodetic_System and www.dqts.net/wgs84.htm websites. See also the website of the *International Earth Rotation Service* at hpiers.obspm.fr. Cited on page 188.

146 G. HECKMAN & M. VAN HAANDEL, *De vele beweijzen van Kepler's wet over ellipsenbanen: een nieuwe voor 'het Boek'?*, Nederlands tijdschrift voor natuurkunde 73, pp. 366–368, November 2007. Cited on page 177.

147 W. K. HARTMAN, R. J. PHILLIPS & G. J. TAYLOR, editors, *Origin of the Moon*, Lunar and Planetary Institute, 1986. Cited on page 191.

148 If you want to read about the motion of the Moon in all its fascinating details, have a look at MARTIN C. GUTZWILLER, *Moon–Earth–Sun: the oldest three body problem*, Reviews of Modern Physics 70, pp. 589–639, 1998. Cited on page 191.

149 DIETRICH NEUMANN, *Physiologische Uhren von Insekten – Zur Ökophysiologie lunarperiodisch kontrollierter Fortpflanzungszeiten*, Naturwissenschaften 82, pp. 310–320, 1995. Cited on page 192.

150 The origin of the duration of the menstrual cycle is not yet settled; however, there are explanations on how it becomes synchronized with other cycles. For a general explanation see

ARKADY PIKOVSKY, MICHAEL ROSENBLUM & JÜRGEN KURTHS, *Synchronization: A Universal Concept in Nonlinear Science*, Cambridge University Press, 2002. Cited on page 192.

151 J. LASKAR, F. JOUTEL & P. ROBUTEL, *Stability of the Earth's obliquity by the moon*, Nature 361, pp. 615–617, 1993. However, the question is not completely settled, and other opinions exist. Cited on page 192.

152 NEIL F. COMINS, *What if the Moon Did not Exist? – Voyages to Earths that Might Have Been*, Harper Collins, 1993. Cited on page 192.

153 A recent proposal is M. ĆUK, D. P. HAMILTON, S. J. LOCK & S. T. STEWART, *Tidal evolution of the Moon from a high-obliquity, high-angular-momentum Earth*, Nature 539, pp. 402–406, 2016. Cited on page 192.

154 M. CONNORS, C. VEILLET, R. BRASSER, P. A. WIEGERT, P. W. CHODAS, S. MIKKOLA & K. A. INNANEN, *Discovery of Earth's quasi-satellite*, Meteoritics & Planetary Science 39, pp. 1251–1255, 2004, and R. BRASSER, K. A. INNANEN, M. CONNORS, C. VEILLET, P. A. WIEGERT, S. MIKKOLA & P. W. CHODAS, *Transient co-orbital asteroids*, Icarus 171, pp. 102–109, 2004. See also the orbits frawn in M. CONNORS, C. VEILLET, R. BRASSER, P. A. WIEGERT, P. W. CHODAS, S. MIKKOLA & K. A. INNANEN, *Horseshoe asteroids and quasi-satellites in Earth-like orbits*, Lunar and Planetary Science 35, p. 1562, 2004,, preprint at www.lpi.usra.edu/meetings/lpsc2004/pdf/1565.pdf. Cited on page 195.

155 P. A. WIEGERT, K. A. INNANEN & S. MIKKOLA, *An asteroidal companion to the Earth*, Nature 387, pp. 685–686, 12 June 1997, together with the comment on pp. 651–652. Details on the orbit and on the fact that Lagrangian points do not always form equilateral triangles can be found in F. NAMOUNI, A. A. CHRISTOU & C. D. MURRAY, *Coorbital dynamics at large eccentricity and inclination*, Physical Review Letters 83, pp. 2506–2509, 1999. Cited on page 194.

156 SIMON NEWCOMB, Astronomical Papers of the American Ephemeris 1, p. 472, 1882. Cited on page 197.

157 For an animation of the tides, have a look at www.jason.oceanobs.com/html/applications/marees/m2_atlantique_fr.html. Cited on page 197.

158 A beautiful introduction is the classic G. FALK & W. RUPPEL, *Mechanik, Relativität, Gravitation – ein Lehrbuch*, Springer Verlag, Dritte Auflage, 1983. Cited on page 197.

159 J. SOLDNER, *Berliner Astronomisches Jahrbuch auf das Jahr 1804*, 1801, p. 161. Cited on page 201.

160 The equality was first tested with precision by R. VON EÖTVÖS, Annalen der Physik & Chemie 59, p. 354, 1896, and by R. VON EÖTVÖS, V. PEKÁR, E. FEKETE, *Beiträge zum Gesetz der Proportionalität von Trägheit und Gravität*, Annalen der Physik 4, Leipzig 68, pp. 11–66, 1922. He found agreement to 5 parts in 10^9. More experiments were performed by P. G. ROLL, R. KROTKOW & R. H. DICKE, *The equivalence of inertial and passive gravitational mass*, Annals of Physics (NY) 26, pp. 442–517, 1964, one of the most interesting and entertaining research articles in experimental physics, and by V. B. BRAGINSKY & V. I. PANOV, Soviet Physics – JETP 34, pp. 463–466, 1971. Modern results, with errors less than one part in 10^{12}, are by Y. SU & al., *New tests of the universality of free fall*, Physical Review D50, pp. 3614–3636, 1994. Several experiments have been proposed to test the equality in space to less than one part in 10^{16}. Cited on page 202.

161 H. EDELMANN, R. NAPIWOTZKI, U. HEBER, N. CHRISTLIEB & D. REIMERS, *HE 0437-5439: an unbound hyper-velocity B-type star*, The Astrophysical Journal 634, pp. L181–

L184, 2005. Cited on page 210.

162 This is explained for example by D.K. Firpić & I.V. Aniçin, *The planets, after all, may run only in perfect circles – but in the velocity space!*, European Journal of Physics 14, pp. 255–258, 1993. Cited on pages 210 and 505.

163 See L. Hodges, *Gravitational field strength inside the Earth*, American Journal of Physics 59, pp. 954–956, 1991. Cited on page 211.

164 The controversial argument is proposed in A. E. Chubykalo & S. J. Vlaev, *Theorem on the proportionality of inertial and gravitational masses in classical mechanics*, European Journal of Physics 19, pp. 1–6, 1998, preprint at arXiv.org/abs/physics/9703031. This paper might be wrong; see the tough comment by B. Jancovici, European Journal of Physics 19, p. 399, 1998, and the reply in arxiv.org/abs/physics/9805003. Cited on page 212.

165 P. Mohazzabi & M. C. James, *Plumb line and the shape of the Earth*, American Journal of Physics 68, pp. 1038–1041, 2000. Cited on page 212.

166 From Neil de Gasse Tyson, *The Universe Down to Earth*, Columbia University Press, 1994. Cited on page 213.

167 G. D. Quinlan, *Planet X: a myth exposed*, Nature 363, pp. 18–19, 1993. Cited on page 213.

168 See en.wikipedia.org/wiki/90377_Sedna. Cited on page 214.

169 See R. Matthews, *Not a snowball's chance ...*, New Scientist 12 July 1997, pp. 24–27. The original claim is by Louis A. Frank, J. B. Sigwarth & J. D. Craven, *On the influx of small comets into the Earth's upper atmosphere*, parts I and II, Geophysical Research Letters 13, pp. 303–306, pp. 307–310, 1986. The latest observations have disproved the claim. Cited on page 215.

170 The ray form is beautifully explained by J. Evans, *The ray form of Newton's law of motion*, American Journal of Physics 61, pp. 347–350, 1993. Cited on page 216.

171 This is a small example from the beautiful text by Mark P. Silverman, *And Yet It Moves: Strange Systems and Subtle Questions in Physics*, Cambridge University Press, 1993. It is a treasure chest for anybody interested in the details of physics. Cited on page 217.

172 G. -L. Lesage, *Lucrèce Newtonien*, Nouveaux mémoires de l'Académie Royale des Sciences et Belles Lettres pp. 404–431, 1747, or www3.bbaw.de/bibliothek/digital/struktur/03-nouv/1782/jpg-0600/00000495.htm. See also en.wikipedia.org/wiki/Le_Sage's_theory_of_gravitation. In fact, the first to propose the idea of gravitation as a result of small particles pushing masses around was Nicolas Fatio de Duillier in 1688. Cited on page 218.

173 J. Laskar, *A numerical experiment on the chaotic behaviour of the solar system*, Nature 338, pp. 237–238, 1989, and J. Laskar, *The chaotic motion of the solar system - A numerical estimate of the size of the chaotic zones*, Icarus 88, pp. 266–291, 1990. The work by Laskar was later expanded by Jack Wisdom, using specially built computers, following only the planets, without taking into account the smaller objects. For more details, see G. J. Sussman & J. Wisdom, *Chaotic Evolution of the Solar System*, Science 257, pp. 56–62, 1992. Today, such calculations can be performed on your home PC with computer code freely available on the internet. Cited on page 219.

174 B. Dubrulle & F. Graner, *Titius-Bode laws in the solar system. 1: Scale invariance explains everything*, Astronomy and Astrophysics 282, pp. 262–268, 1994, and *Titius-Bode laws in the solar system. 2: Build your own law from disk models*, Astronomy and Astrophysics 282, pp. 269--276, 1994. Cited on page 221.

175 M. Lecar, *Bode's Law*, Nature 242, pp. 318–319, 1973, and M. Henon, *A comment on "The resonant structure of the solar system" by A.M. Molchanov*, Icarus 11, pp. 93–94, 1969. Cited on page 221.

176 CASSIUS DIO, *Historia Romana, c.* 220, book 37, 18. For an English translation, see the site penelope.uchicago.edu/Thayer/E/Roman/Texts/Cassius_Dio/37*.html. Cited on page 221.

177 See the beautiful paper A. J. SIMOSON, *Falling down a hole through the Earth*, Mathematics Magazine 77, pp. 171–188, 2004. See also A. J. SIMOSON, *The gravity of Hades*, 75, pp. 335–350, 2002. Cited on pages 223 and 507.

178 M. BEVIS, D. ALSDORF, E. KENDRICK, L. P. FORTES, B. FORSBERG, R. MALLEY & J. BECKER, *Seasonal fluctuations in the mass of the Amazon River system and Earth's elastic response*, Geophysical Research Letters 32, p. L16308, 2005. Cited on page 223.

179 D. HESTENES, M. WELLS & G. SWACKHAMER, *Force concept inventory*, Physics Teacher 30, pp. 141–158, 1982. The authors developed tests to check the understanding of the concept of physical force in students; the work has attracted a lot of attention in the field of physics teaching. Cited on page 229.

180 For a general overview on friction, from physics to economics, architecture and organizational theory, see N. ÅKERMAN, editor, *The Necessity of Friction – Nineteen Essays on a Vital Force*, Springer Verlag, 1993. Cited on page 234.

181 See M. HIRANO, K. SHINJO, R. KANECKO & Y. MURATA, *Observation of superlubricity by scanning tunneling microscopy*, Physical Review Letters 78, pp. 1448–1451, 1997. See also the discussion of their results by SERGE FAYEULLE, *Superlubricity: when friction stops*, Physics World pp. 29–30, May 1997. Cited on page 234.

182 C. DONALD AHRENS, *Meteorology Today: An Introduction to the Weather, Climate, and the Environment*, West Publishing Company, 1991. Cited on page 235.

183 This topic is discussed with lucidity by J. R. MUREIKA, *What really are the best 100 m performances?*, Athletics: Canada's National Track and Field Running Magazine, July 1997. It can also be found as arxiv.org/abs/physics/9705004, together with other papers on similar topics by the same author. Cited on page 235.

184 F. P. BOWDEN & D. TABOR, *The Friction and Lubrication of Solids*, Oxford University Press, Part I, revised edition, 1954, and part II, 1964. Cited on page 236.

185 A powerful book on human violence is JAMES GILLIGAN, *Violence – Our Deadly Epidemic and its Causes*, Grosset/Putnam, 1992. Cited on page 236.

186 The main tests of randomness of number series – among them the gorilla test – can be found in the authoritative paper by G. MARSAGLIA & W. W. TSANG, *Some difficult-to-pass tests of randomness*, Journal of Statistical Software 7, p. 8, 2002. It can also be downloaded from www.jstatsoft.org/v07/i03. Cited on page 239.

187 See the interesting book on the topic by JAROSLAW STRZALKO, JULIUSZ GRABSKI, PRZEMYSLAW PERLIKOWSKI, ANDRZEIJ STEFANSKI & TOMASZ KAPITANIAK, *Dynamics of Gambling: Origin of Randomness in Mechanical Systems*, Springer, 2009, as well as the more recent publications by Kapitaniak. Cited on page 239.

188 For one aspect on free will, see the captivating book by BERT HELLINGER, *Zweierlei Glück*, Carl Auer Systeme Verlag, 1997. The author explains how to live serenely and with the highest possible responsibility for one's actions, by reducing entanglements with the destiny of others. He describes a powerful technique to realise this goal.

A completely different viewpoint is given by AUNG SAN SUU KYI, *Freedom from Fear*, Penguin, 1991. One of the bravest women on Earth, she won the Nobel Peace Price in 1991.

An effective personal technique is presented by PHIL STUTZ & BARRY MICHELS, *The Tools*, Spiegel & Grau, 2012. Cited on page 242.

189 HENRIK WALTER, *Neurophilosophie der Willensfreiheit*, Mentis Verlag, Paderborn 1999. Also available in English translation. Cited on page 242.

190 Giuseppe Fumagalli, *Chi l'ha detto?*, Hoepli, 1983. Cited on page 242.

191 See the tutorial on the Peaucellier-Lipkin linkage by D.W. Henderson and D. Taimina found on kmoddl.library.cornell.edu/tutorials/11/index.php. The internet contains many other pages on the topic. Cited on page 243.

192 The beautiful story of the south-pointing carriage is told in Appendix B of James Foster & J. D. Nightingale, *A Short Course in General Relativity*, Springer Verlag, 2nd edition, 1998. Such carriages have existed in China, as told by the great sinologist Joseph Needham, but their construction is unknown. The carriage described by Foster and Nightingale is the one reconstructed in 1947 by George Lancaster, a British engineer. Cited on page 244.

193 T. Van de Kamp, P. Vagovic, T. Baumbach & A. Riedel, *A biological screw in a beetle's leg*, Science 333, p. 52, 2011. Cited on page 244.

194 M. Burrows & G. P. Sutton, *Interacting gears synchronise propulsive leg movements in a jumping insect*, Science 341, pp. 1254–1256, 2013. Cited on page 244.

195 See for example Z. Ghahramani, *Building blocks of movement*, Nature 407, pp. 682–683, 2000. Researchers in robot control are also interested in such topics. Cited on page 244.

196 G. Gutierrez, C. Fehr, A. Calzadilla & D. Figueroa, *Fluid flow up the wall of a spinning egg*, American Journal of Physics 66, pp. 442–445, 1998. Cited on page 244.

197 A historical account is given in Wolfgang Yourgray & Stanley Mandelstam, *Variational Principles in Dynamics and Quantum Theory*, Dover, 1968. Cited on pages 248 and 256.

198 C. G. Gray & E. F. Taylor, *When action is not least*, American Journal of Physics 75, pp. 434–458, 2007. Cited on page 253.

199 Max Päsler, *Prinzipe der Mechanik*, Walter de Gruyter & Co., 1968. Cited on page 254.

200 The relations between possible Lagrangians are explained by Herbert Goldstein, *Classical Mechanics*, 2nd edition, Addison-Wesley, 1980. Cited on page 255.

201 The Hemingway statement is quoted by Marlene Dietrich in Aaron E. Hotchner, *Papa Hemingway*, Random House, 1966, in part 1, chapter 1. Cited on page 256.

202 C. G. Gray, G. Karl & V. A. Novikov, *From Maupertius to Schrödinger. Quantization of classical variational principles*, American Journal of Physics 67, pp. 959–961, 1999. Cited on page 257.

203 J. A. Moore, *An innovation in physics instruction for nonscience majors*, American Journal of Physics 46, pp. 607–612, 1978. Cited on page 257.

204 See e.g. Alan P. Boss, *Extrasolar planets*, Physics Today 49, pp. 32–38. September 1996. The most recent information can be found at the 'Extrasolar Planet Encyclopaedia' maintained at www.obspm.fr/planets by Jean Schneider at the Observatoire de Paris. Cited on page 261.

205 A good review article is by David W. Hughes, *Comets and Asteroids*, Contemporary Physics 35, pp. 75–93, 1994. Cited on page 261.

206 G. B. West, J. H. Brown & B. J. Enquist, *A general model for the origin of allometric scaling laws in biology*, Science 276, pp. 122–126, 4 April 1997, with a comment on page 34 of the same issue. The rules governing branching properties of blood vessels, of lymph systems and of vessel systems in plants are explained. For more about plants, see also the paper G. B. West, J. H. Brown & B. J. Enquist, *A general model for the structure and allometry of plant vascular systems*, Nature 400, pp. 664–667, 1999. Cited on page 263.

207 J. R. Banavar, A. Martin & A. Rinaldo, *Size and form in efficient transportation networks*, Nature 399, pp. 130–132, 1999. Cited on page 263.

208 N. MOREIRA, *New striders - new humanoids with efficient gaits change the robotics land-scape*, Science News Online 6th of August, 2005. Cited on page 264.

209 WERNER HEISENBERG, *Der Teil und das Ganze*, Piper, 1969. Cited on page 266.

210 See the clear presenttion by E. H. LOCKWOOD & R. H. MACMILLAN, *Geometric Symmetry*, Cambridge University Press, 1978. Cited on page 266.

211 JOHN MANSLEY ROBINSON, *An Introduction to Early Greek Philosophy*, Houghton Muffin 1968, chapter 5. Cited on page 270.

212 See e.g. B. BOWER, *A child's theory of mind*, Science News 144, pp. 40–41. Cited on page 271.

213 The most beautiful book on this topic is the text by BRANKO GRÜNBAUM & G. C. SHEPHARD, *Tilings and Patterns*, W.H. Freeman and Company, New York, 1987. It has been translated into several languages and republished several times. Cited on page 273.

214 About tensors and ellipsoids in three-dimensional space, see mysite.du.edu/~jcalvert/phys/ellipso.htm. In four-dimensional space-time, tensors are more abstract to comprehend. With emphasis on their applications in relativity, such tensors are explained in R. FROSCH, *Four-tensors, the mother tongue of classical physics*, vdf Hochschulverlag, 2006, partly available on books.google.com. Cited on page 278.

215 U. NIEDERER, *The maximal kinematical invariance group of the free Schrödinger equation*, Helvetica Physica Acta 45, pp. 802–810, 1972. See also the introduction by O. JAHN & V. V. SREEDHAR, *The maximal invariance group of Newton's equations for a free point particle*, arxiv.org/abs/math-ph/0102011. Cited on page 279.

216 The story is told in the interesting biography of Einstein by A. PAIS, *'Subtle is the Lord...'* – *The Science and the Life of Albert Einstein*, Oxford University Press, 1982. Cited on page 280.

217 W. ZÜRN & R. WIDMER-SCHNIDRIG, *Globale Eigenschwingungen der Erde*, Physik Journal 1, pp. 49–55, 2002. Cited on page 290.

218 N. GAUTHIER, *What happens to energy and momentum when two oppositely-moving wave pulses overlap?*, American Journal of Physics 71, pp. 787–790, 2003. Cited on page 300.

219 An informative and modern summary of present research about the ear and the details of its function is www.physicsweb.org/article/world/15/5/8. Cited on page 303.

220 A renowned expert of the physics of singing is Ingo Titze. Among his many books and papers is the popular introduction I. R. TITZE, *The human instrument*, Scientific American pp. 94–101, January 2008. Several of his books, papers and presentation are free to download on the website www.ncvs.org of the National Center of Voice & Speech. They are valuable to everybody who has a passion for singing and the human voice. See also the article and sound clips at www.scientificamerican.com/article.cfm?id=sound-clips-human-instrument. An interesting paper is also M. KOB & al., *Analysing and understanding the singing voice: recent progress and open questions*, Current Bioinformatics 6, pp. 362–374, 2011. Cited on page 308.

221 S. ADACHI, *Principles of sound production in wind instruments*, Acoustical Science and Technology 25, pp. 400–404, 2004. Cited on page 309.

222 The literature on tones and their effects is vast. For example, people have explored the differences and effects of various intonations in great detail. Several websites, such as bellsouthpwp.net/j/d/jdelaub/jstudio.htm, allow listening to music played with different intonations. People have even researched whether animals use just or chromatic intonation.

(See, for example, K. Leutwyler, *Exploring the musical brain*, Scientific American January 2001.) There are also studies of the effects of low frequencies, of beat notes, and of many other effects on humans. However, many studies mix serious and non-serious arguments. It is easy to get lost in them. Cited on page 310.

223 M. Fatemi, P. L. Ogburn & J. F. Greenleaf, *Fetal stimulation by pulsed diagnostic ultrasound*, Journal of Ultrasound in Medicine 20, pp. 883–889, 2001. See also M. Fatemi, A. Alizad & J. F. Greenleaf, *Characteristics of the audio sound generated by ultrasound imaging systems*, Journal of the Acoustical Society of America 117, pp. 1448–1455, 2005. Cited on page 313.

224 I know a female gynecologist who, during her own pregnancy, imaged her child *every day* with her ultrasound machine. The baby was born with strong hearing problems that did not go away. Cited on page 313.

225 R. Mole, *Possible hazards of imaging and Doppler ultrasound in obstetrics*, Birth Issues in Perinatal Care 13, pp. 29–37, 2007. Cited on page 314.

226 A. L. Hodgkin & A. F. Huxley, *A quantitative description of membrane current and its application to conduction and excitation in nerve*, Journal of Physiology 117, pp. 500–544, 1952. This famous paper of theoretical biology earned the authors the Nobel Prize in Medicine in 1963. Cited on page 315.

227 T. Filippov, *The Versatile Soliton*, Springer Verlag, 2000. See also J. S. Russel, *Report of the Fourteenth Meeting of the British Association for the Advancement of Science*, Murray, London, 1844, pp. 311–390. Cited on pages 316 and 318.

228 R. S. Ward, *Solitons and other extended field configurations*, preprint at arxiv.org/abs/hep-th/0505135. Cited on page 318.

229 D. B. Bahr, W. T. Pfeffer & R. C. Browning, *The surprising motion of ski moguls*, Physics Today 62, pp. 68–69, November 2009. Cited on page 319.

230 N. J. Zabusky & M. D. Kruskal, *Interaction of solitons in a collisionless plasma and the recurrence of initial states*, Physical Review Letters 15, pp. 240–243, 1965. Cited on page 317.

231 O. Muskens, *De kortste knal ter wereld*, Nederlands tijdschrift voor natuurkunde pp. 70–73, 2005. Cited on page 318.

232 E. Heller, *Freak waves: just bad luck, or avoidable?*, Europhysics News pp. 159–161, September/October 2005, downloadable at www.europhysicsnews.org. Cited on page 320.

233 See the beautiful article by D. Aarts, M. Schmidt & H. Lekkerkerker, *Directe visuele waarneming van thermische capillaire golven*, Nederlands tijdschrift voor natuurkunde 70, pp. 216–218, 2004. Cited on page 321.

234 For more about the ocean sound channel, see the novel by Tom Clancy, *The Hunt for Red October*. See also the physics script by R. A. Muller, *Government secrets of the oceans, atmosphere, and UFOs*, web.archive.org/web/*/muller.lbl.gov/teaching/Physics10/chapters/9-SecretsofUFOs.html 2001. Cited on page 322.

235 B. Wilson, R. S. Batty & L. M. Dill, *Pacific and Atlantic herring produce burst pulse sounds*, Biology Letters 271, number S3, 7 February 2004. Cited on page 323.

236 A. Chabchoub & M. Fink, *Time-reversal generation of rogue Wwaves*, Physical Review Letters 112, p. 124101, 2014, preprint at arxiv.org/abs/1311.2990. See also the cited references. Cited on page 324.

237 See for example the article by G. Fritsch, *Infraschall*, Physik in unserer Zeit 13, pp. 104–110, 1982. Cited on page 325.

238 Wavelet transformations were developed by the French mathematicians Alex Grossmann, Jean Morlet and Thierry Paul. The basic paper is A. GROSSMANN, J. MORLET & T. PAUL, *Integral transforms associated to square integrable representations*, Journal of Mathematical Physics 26, pp. 2473–2479, 1985. For a modern introduction, see STÉPHANE MALLAT, *A Wavelet Tour of Signal Processing*, Academic Press, 1999. Cited on page 326.

239 P. MANOGG, *Knall und Superknall beim Überschallflug*, Der mathematische und naturwissenschaftliche Unterricht 35, pp. 26–33, 1982, my physics teacher in secondary school. Cited on page 326.

240 See the excellent introduction by L. ELLERBREOK & L. VAN DEN HOORN, *In het kielzog van Kelvin*, Nederlands tijdschrift voor natuurkunde 73, pp. 310–313, 2007. About exceptions to the Kelvin angle, see www.graingerdesigns.net/oshunpro/design-technology/wave-cancellation. Cited on page 327.

241 JAY INGRAM, *The Velocity of Honey - And More Science of Everyday Life*, Viking, 2003. See also W. W. L. AU & J. A. SIMMONS, *Echolocation in dolphins and bats*, Physics Today 60, pp. 40–45, 2007. Cited on page 328.

242 M. BOITI, J.-P. LEON, L. MARTINA & F. PEMPINELLI, *Scattering of localized solitons in the plane*, Physics Letters A 132, pp. 432–439, 1988, A. S. FOKAS & P. M. SANTINI, *Coherent structures in multidimensions*, Physics Review Letters 63, pp. 1329–1333, 1989, J. HIETARINTA & R. HIROTA, *Multidromion solutions to the Davey–Stewartson equation*, Physics Letters A 145, pp. 237–244, 1990. Cited on page 328.

243 For some of this fascinating research, see J. L. HAMMACK, D. M. HENDERSON & H. SEGUR, *Progressive waves with persistent two-dimensional surface patterns in deep water*, Journal of Fluid Mechanics 532, pp. 1–52, 2005. For a beautiful photograph of crossing cnoidal waves, see A. R. OSBORNE, *Hyperfast Modeling of Shallow-Water Waves: The KdV and KP Equations*, International Geophysics 97, pp. 821–856, 2010. See also en.wikipedia.org/wiki/Waves_and_shallow_water, en.wikipedia.org/wiki/Cnoidal_wave and en.wikipedia.org/wiki/Tidal_bore for a first impression. Cited on page 329.

244 The sound frequency change with bottle volume is explained on hyperphysics.phy-astr.gsu.edu/Hbase/Waves/cavity.html. Cited on page 329.

245 A passionate introduction is NEVILLE H. FLETCHER & THOMAS D. ROSSING, *The Physics of Musical Instruments*, second edition, Springer 2000. Cited on page 330.

246 M. AUSLOOS & D. H. BERMAN, *Multivariate Weierstrass–Mandelbrot function*, Proceedings of the Royal Society in London A 400, pp. 331–350, 1985. Cited on page 333.

247 Catechism of the Catholic Church, Part Two, Section Two, Chapter One, Article 3, statements 1376, 1377 and 1413, found at www.vatican.va/archive/ENG0015/__P4I.HTM or www.vatican.va/archive/ITA0014/__P40.HTM with their explanations on www.vatican.va/archive/compendium_ccc/documents/archive_2005_compendium-ccc_en.html and www.vatican.va/archive/compendium_ccc/documents/archive_2005_compendium-ccc_it.html. Cited on page 336.

248 The original text of the 1633 conviction of Galileo can be found on it.wikisource.org/wiki/Sentenza_di_condanna_di_Galileo_Galilei. Cited on page 336.

249 The retraction that Galileo was forced to sign in 1633 can be found on it.wikisource.org/wiki/Abiura_di_Galileo_Galilei. Cited on page 336.

250 M. ARTIGAS, *Un nuovo documento sul caso Galileo: EE 291*, Acta Philosophica 10, pp. 199–214, 2001. Cited on page 337.

251 Most of these points are made, directly or indirectly, in the book by ANNIBALE FANTOLI, *Galileo: For Copernicanism and for the Church*, Vatican Observatory Publications, second edition, 1996, and by George Coyne, director of the Vatican observatory, in his speeches and publications, for example in G. Coyne, *Galileo: for Copernicanism and for the church*, Zwoje 3/36, 2003, found at www.zwoje-scrolls.com/zwoje36/text05p.htm. Cited on page 337.

252 Thomas A. McMahon & John T. Bonner, *On Size and Life*, Scientific American/Freeman, 1983. Another book by John Bonner, who won the Nobel Prize in Biology, is John T. Bonner, *Why Size Matters: From Bacteria to Blue Whales*, Princeton University Press, 2011. Cited on page 337.

253 G. W. Koch, S. C. Sillett, G. M. Jennings & S. D. Davis, *The limits to tree height*, Nature 428, pp. 851–854, 2004. Cited on page 338.

254 A simple article explaining the tallness of trees is A. Mineyev, *Trees worthy of Paul Bunyan*, Quantum pp. 4–10, January–February 1994. (Paul Bunyan is a mythical giant lumberjack who is the hero of the early frontier pioneers in the United States.) Note that the transport of liquids in trees sets no limits on their height, since water is pumped up along tree stems (except in spring, when it is pumped up from the roots) by evaporation from the leaves. This works almost without limits because water columns, when nucleation is carefully avoided, can be put under tensile stresses of over 100 bar, corresponding to 1000 m. See also P. Nobel, *Plant Physiology*, Academic Press, 2nd Edition, 1999. An artificial tree – though extremely small – using the same mechanism was built and studied by T. D. Wheeler & A. D. Stroock, *The transpiration of water at negative pressures in a synthetic tree*, Nature 455, pp. 208–212, 2008. See also N. M. Holbrook & M. A. Zwieniecki, *Transporting water to the top of trees*, Physics Today pp. 76–77, 2008. Cited on pages 338 and 359.

255 Such information can be taken from the excellent overview article by M. F. Ashby, *On the engineering properties of materials*, Acta Metallurgica 37, pp. 1273–1293, 1989. The article explains the various general criteria which determine the selection of materials, and gives numerous tables to guide the selection. Cited on page 338.

256 See the beautiful paper by S. E. Virgo, *Loschmidt's number*, Science Progress 27, pp. 634–649, 1933. It is also freely available in HTML format on the internet. Cited on pages 340 and 342.

257 See the delightful paper by Peter Pesic, *Estimating Avogadro's number from skylight and airlight*, European Journal of Physics 26, pp. 183–187, 2005. The mistaken statement that the blue colour is due to density fluctuations is dispelled in C. F. Bohren & A. B. Fraser, *Color of the Sky*, Physics Teacher 238, pp. 267–272, 1985. It also explains that the variations of the sky colour, like the colour of milk, are due to multiple scattering. Cited on pages 341 and 515.

258 For a photograph of a single barium *atom* – named Astrid – see Hans Dehmelt, *Experiments with an isolated subatomic particle at rest*, Reviews of Modern Physics 62, pp. 525–530, 1990. For another photograph of a barium *ion*, see W. Neuhauser, M. Hohenstatt, P. E. Toschek & H. Dehmelt, *Localized visible Ba⁺ mono-ion oscillator*, Physical Review A 22, pp. 1137–1140, 1980. See also the photograph on page 344. Cited on page 344.

259 Holograms of atoms were first produced by Hans-Werner Fink & al., *Atomic resolution in lens-less low-energy electron holography*, Physical Review Letters 67, pp. 1543–1546, 1991. Cited on page 344.

260 A single–atom laser was built in 1994 by K. An, J. J. Childs, R. R. Dasari & M. S. Feld, *Microlaser: a laser with one atom in an optical resonator*, Physical Review Letters 73, p. 3375, 1994. Cited on page 344.

261 The photograph on the left of Figure 240 on page 344 is the first image that showed sub-atomic structures (visible as shadows on the atoms). It was published by F. J. GIESSIBL, S. HEMBACHER, H. BIELEFELDT & J. MANNHART, *Subatomic features on the silicon (111)-(7x7) surface observed by atomic force microscopy*, Science **289**, pp. 422 – 425, 2000. Cited on page 344.

262 See for example C. SCHILLER, A. A. KOOMANS, T. L. VAN ROOY, C. SCHÖNENBERGER & H. B. ELSWIJK, *Decapitation of tungsten field emitter tips during sputter sharpening*, Surface Science Letters **339**, pp. L925–L930, 1996. Cited on page 344.

263 U. WEIERSTALL & J. C. H. SPENCE, *An STM with time-of-flight analyzer for atomic species identification*, MSA 2000, Philadelphia, Microscopy and Microanalysis **6**, Supplement 2, p. 718, 2000. Cited on page 345.

264 P. KREHL, S. ENGEMANN & D. SCHWENKEL, *The puzzle of whip cracking – uncovered by a correlation of whip-tip kinematics with shock wave emission*, Shock Waves **8**, pp. 1–9, 1998. The authors used high-speed cameras to study the motion of the whip. A new aspect has been added by A. GORIELY & T. MCMILLEN, *Shape of a cracking whip*, Physical Review Letters **88**, p. 244301, 2002. This article focuses on the tapered shape of the whip. However, the neglection of the tuft – a piece at the end of the whip which is required to make it crack – in the latter paper shows that there is more to be discovered still. Cited on page 349.

265 Z. SHENG & K. YAMAFUJI, *Realization of a Human Riding a Unicycle by a Robot*, Proceedings of the 1995 IEEE International Conference on Robotics and Automation, Vol. 2, pp. 1319–1326, 1995. Cited on page 349.

266 On human unicycling, see JACK WILEY, *The Complete Book of Unicycling*, Lodi, 1984, and SEBASTIAN HOEHER, *Einradfahren und die Physik*, Reinbeck, 1991. Cited on page 349.

267 W. THOMSON, *Lecture to the Royal Society of Edinburgh*, 18 February 1867, Proceedings of the Royal Society in Edinborough **6**, p. 94, 1869. Cited on page 350.

268 S. T. THORODDSEN & A. Q. SHEN, *Granular jets*, Physics of Fluids **13**, pp. 4–6, 2001, and A. Q. SHEN & S. T. THORODDSEN, *Granular jetting*, Physics of Fluids **14**, p. S3, 2002, Cited on page 350.

269 M. J. HANCOCK & J. W. M. BUSH, *Fluid pipes*, Journal of Fluid Mechanics **466**, pp. 285–304, 2002. A. E. HOSOI & J. W. M. BUSH, *Evaporative instabilities in climbing films*, Journal of Fluid Mechanics **442**, pp. 217–239, 2001. J. W. M. BUSH & A. E. HASHA, *On the collision of laminar jets: fluid chains and fishbones*, Journal of Fluid Mechanics **511**, pp. 285–310, 2004. Cited on page 354.

270 The present record for negative pressure in water was achieved by Q. ZHENG, D. J. DURBEN, G. H. WOLF & C. A. ANGELL, *Liquids at large negative pressures: water at the homogeneous nucleation limit*, Science **254**, pp. 829–832, 1991. Cited on page 359.

271 H. MARIS & S. BALIBAR, *Negative pressures and cavitation in liquid helium*, Physics Today **53**, pp. 29–34, 2000. Sebastien Balibar has also written several popular books that are presented at his website www.lps.ens.fr/~balibar. Cited on page 359.

272 The present state of our understanding of turbulence is described by G. FALKOVICH & K. P. SREENIVASAN, *Lessons from hydrodynamic turbulence*, Physics Today **59**, pp. 43–49, 2006. Cited on page 363.

273 K. WELTNER, *A comparison of explanations of aerodynamical lifting force*, American Journal of Physics **55**, pp. 50–54, 1987, K. WELTNER, *Aerodynamic lifting force*, The Physics Teacher **28**, pp. 78–82, 1990. See also the user.uni-frankfurt.de/~weltner/Flight/PHYSIC4.htm and the www.av8n.com/how/htm/airfoils.html websites. Cited on page 363.

274 L. LANOTTE,   J. MAUER,   S. MENDEZ,   D. A. FEDOSOV,   J.-M. FROMENTAL, V. CLAVERIA, F. NICOUD, G. GOMPPER & M. ABKARIAN, *Red cells' dynamic morphologies govern blood shear thinning under microcirculatory flow conditions*, Proceedings of the National Academy of Sciences of the United States of America 2016, preprint at arxiv.org/abs/1608.03730. Cited on page 368.

275 S. GEKLE, I. R. PETERS, J. M. GORDILLO, D. van der MEER & D. LOHSE, *Supersonic air flow due to solid-liquid impact*, Physical Review Letters 104, p. 024501, 2010. Films of the effect can be found at physics.aps.org/articles/v3/4. Cited on page 370.

276 See the beautiful book by RAINER F. FOELIX, *Biologie der Spinnen*, Thieme Verlag, 1996, also available in an even newer edition in English as RAINER F. FOELIX, *Biology of Spiders*, Oxford University Press, third edition, 2011. Special fora dedicated only to spiders can be found on the internet. Cited on page 370.

277 See the website www.esa.int/esaCP/SEMER89U7TG_index_0.html. Cited on page 370.

278 For a fascinating account of the passion and the techniques of apnoea diving, see UMBERTO PELIZZARI, *L'Homme et la mer*, Flammarion, 1994. Palizzari cites and explains the saying by Enzo Maiorca: 'The first breath you take when you come back to the surface is like the first breath with which you enter life.' Cited on page 371.

279 LYDÉRIC BOCQUET, *The physics of stone skipping*, American Journal of Physics 17, pp. 150–155, 2003. The present recod holder is Kurt Steiner, with 40 skips. See pastoneskipping.com/steiner.htm and www.stoneskipping.com. The site www.yeeha.net/nassa/guin/g2.html is by the a previous world record holder, Jerdome Coleman-McGhee. Cited on page 374.

280 S. F. KISTLER & L. E. SCRIVEN, *The teapot effect: sheetforming flows with deflection, wetting, and hysteresis*, Journal of Fluid Mechanics 263, pp. 19–62, 1994. Cited on page 376.

281 J. WALKER, *Boiling and the Leidenfrost effect*, a chapter from DAVID HALLIDAY, ROBERT RESNICK & JEARL WALKER, *Fundamentals of Physics*, Wiley, 2007. The chapter can also be found on the internet as pdf file. Cited on page 377.

282 E. HOLLANDER, *Over trechters en zo ...*, Nederlands tijdschrift voor natuurkunde 68, p. 303, 2002. Cited on page 378.

283 S. DORBOLO, H. CAPS & N. VANDEWALLE, *Fluid instabilities in the birth and death of antibubbles*, New Journal of Physics 5, p. 161, 2003. Cited on page 379.

284 T. T. LIM, *A note on the leapfrogging between two coaxial vortex rings at low Reynolds numbers*, Physics of Fluids 9, pp. 239–241, 1997. Cited on page 380.

285 P. AUSSILLOUS & D. QUÉRÉ, *Properties of liquid marbles*, Proc. Roy. Soc. London 462, pp. 973–999, 2006, and references therein. Cited on page 380.

286 Thermostatics and thermodynamics are difficult to learn also because the fields were not discovered in a systematic way. See C. TRUESDELL, *The Tragicomical History of Thermodynamics 1822–1854*, Springer Verlag, 1980. An excellent advanced textbook on thermostatics and thermodynamics is LINDA REICHL, *A Modern Course in Statistical Physics*, Wiley, 2nd edition, 1998. Cited on page 383.

287 Gas expansion was the main method used for the definition of the official temperature scale. Only in 1990 were other methods introduced officially, such as total radiation thermometry (in the range 140 K to 373 K), noise thermometry (2 K to 4 K and 900 K to 1235 K), acoustical thermometry (around 303 K), magnetic thermometry (0.5 K to 2.6 K) and optical radiation thermometry (above 730 K). Radiation thermometry is still the central method in the range from about 3 K to about 1000 K. This is explained in detail in R. L. RUSBY, R. P. HUDSON, M. DURIEUX, J. F. SCHOOLEY, P. P. M. STEUR & C. A. SWENSON,

The basis of the ITS-90, Metrologia 28, pp. 9–18, 1991. On the water boiling point see also Ref. 315. Cited on pages 385, 551, and 555.

288 Other methods to rig lottery draws made use of balls of different mass or of balls that are more polished. One example of such a scam was uncovered in 1999. Cited on page 384.

289 See for example the captivating text by GINO SEGRÈ, *A Matter of Degrees: What Temperature Reveals About the Past and Future of Our Species, Planet and Universe*, Viking, New York, 2002. Cited on page 385.

290 D. KARSTÄDT, F. PINNO, K.-P. MÖLLMANN & M. VOLLMER, *Anschauliche Wärmelehre im Unterricht: ein Beitrag zur Visualisierung thermischer Vorgänge*, Praxis der Naturwissenschaften Physik 5-48, pp. 24–31, 1999, K.-P. MÖLLMANN & M. VOLLMER, *Eine etwas andere, physikalische Sehweise - Visualisierung von Energieumwandlungen und Strahlungsphysik für die (Hochschul-)lehre*, Physikalische Blllätter 56, pp. 65–69, 2000, D. KARSTÄDT, K.-P. MÖLLMANN, F. PINNO & M. VOLLMER, *There is more to see than eyes can detect: visualization of energy transfer processes and the laws of radiation for physics education*, The Physics Teacher 39, pp. 371–376, 2001, K.-P. MÖLLMANN & M. VOLLMER, *Infrared thermal imaging as a tool in university physics education*, European Journal of Physics 28, pp. S37–S50, 2007. Cited on page 387.

291 See for example the article by H. PRESTON-THOMAS, *The international temperature scale of 1990 (ITS-90)*, Metrologia 27, pp. 3–10, 1990, and the errata H. PRESTON-THOMAS, *The international temperature scale of 1990 (ITS-90)*, Metrologia 27, p. 107, 1990, Cited on page 391.

292 For an overview, see CHRISTIAN ENSS & SIEGFRIED HUNKLINGER, *Low-Temperature Physics*, Springer, 2005. Cited on page 391.

293 The famous paper on Brownian motion which contributed so much to Einstein's fame is A. EINSTEIN, *Über die von der molekularkinetischen Theorie der Wärme geforderte Bewegung von in ruhenden Flüssigkeiten suspendierten Teilchen*, Annalen der Physik 17, pp. 549–560, 1905. In the following years, Einstein wrote a series of further papers elaborating on this topic. For example, he published his 1905 Ph.D. thesis as A. EINSTEIN, *Eine neue Bestimmung der Moleküldimensionen*, Annalen der Physik 19, pp. 289–306, 1906, and he corrected a small mistake in A. EINSTEIN, *Berichtigung zu meiner Arbeit: 'Eine neue Bestimmung der Moleküldimensionen'*, Annalen der Physik 34, pp. 591–592, 1911, where, using new data, he found the value $6.6 \cdot 10^{23}$ for Avogadro's number. However, five years before Smoluchowski and Einstein, a much more practically-minded man had made the *same* calculations, but in a different domain: the mathematician Louis Bachelier did so in his PhD about stock options; this young financial analyst was thus smarter than Einstein. Cited on page 392.

294 The first experimental confirmation of the prediction was performed by J. PERRIN, Comptes Rendus de l'Académie des Sciences 147, pp. 475–476, and pp. 530–532, 1908. He masterfully sums up the whole discussion in JEAN PERRIN, *Les atomes*, Librarie Félix Alcan, Paris, 1913. Cited on page 394.

295 PIERRE GASPARD & al., *Experimental evidence for microscopic chaos*, Nature 394, p. 865, 27 August 1998. Cited on page 394.

296 An excellent introduction into the physics of heat is the book by LINDA REICHL, *A Modern Course in Statistical Physics*, Wiley, 2nd edition, 1998. Cited on page 395.

297 On the bicycle speed record, see the website fredrompelberg.com. It also shows details of the bicycle he used. On the drag effect of a motorbike behind a bicycle, see B. BLOCKEN, Y. TOPARLAR & T. ANDRIANNE, *Aerodynamic benefit for a cyclist by a following motorcycle*, Journal of Wind Engineering & Industrial Aerodynamics 2016, available for free download at www.urbanphysics.net. Cited on page 381.

298 F. HERRMANN, *Mengenartige Größen im Physikunterricht*, Physikalische Blätter 54, pp. 830–832, September 1998. See also his lecture notes on general introductory physics on the website www.physikdidaktik.uni-karlsruhe.de/skripten. Cited on pages 232, 354, and 395.

299 These points are made clearly and forcibly, as is his style, by N.G. VAN KAMPEN, *Entropie*, Nederlands tijdschrift voor natuurkunde 62, pp. 395–396, 3 December 1996. Cited on page 398.

300 This is a disappointing result of all efforts so far, as Grégoire Nicolis always stresses in his university courses. Seth Lloyd has compiled a list of 31 proposed definitions of complexity, containing among others, fractal dimension, grammatical complexity, computational complexity, thermodynamic depth. See, for example, a short summary in Scientific American p. 77, June 1995. Cited on page 398.

301 Minimal entropy is discussed by L. SZILARD, *Über die Entropieverminderung in einem thermodynamischen System bei Eingriffen intelligenter Wesen*, Zeitschrift für Physik 53, pp. 840–856, 1929. This classic paper can also be found in English translation in his collected works. Cited on page 399.

302 G. COHEN-TANNOUDJI, *Les constantes universelles*, Pluriel, Hachette, 1998. See also L. BRILLOUIN, *Science and Information Theory*, Academic Press, 1962. Cited on page 399.

303 H. W. ZIMMERMANN, *Particle entropies and entropy quanta IV: the ideal gas, the second law of thermodynamics, and the P-t uncertainty relation*, Zeitschrift für physikalische Chemie 217, pp. 55–78, 2003, and H. W. ZIMMERMANN, *Particle entropies and entropy quanta V: the P-t uncertainty relation*, Zeitschrift für physikalische Chemie 217, pp. 1097–1108, 2003. Cited on pages 399 and 400.

304 See for example A. E. SHALYT-MARGOLIN & A. YA. TREGUBOVICH, *Generalized uncertainty relation in thermodynamics*, arxiv.org/abs/gr-qc/0307018, or J. UFFINK & J. VAN LITH-VAN DIS, *Thermodynamic uncertainty relations*, Foundations of Physics 29, p. 655, 1999. Cited on page 399.

305 B. LAVENDA, *Statistical Physics: A Probabilistic Approach*, Wiley-Interscience, 1991. Cited on pages 399 and 400.

306 The quote given is found in the introduction by George Wald to the text by LAWRENCE J. HENDERSON, *The Fitness of the Environment*, Macmillan, New York, 1913, reprinted 1958. Cited on page 400.

307 A fascinating introduction to chemistry is the text by JOHN EMSLEY, *Molecules at an Exhibition*, Oxford University Press, 1998. Cited on page 401.

308 B. POLSTER, *What is the best way to lace your shoes?*, Nature 420, p. 476, 5 December 2002. Cited on page 402.

309 L. BOLTZMANN, *Über die mechanische Bedeutung des zweiten Hauptsatzes der Wärmetheorie*, Sitzungsberichte der königlichen Akademie der Wissenschaften in Wien 53, pp. 155–220, 1866. Cited on page 403.

310 See for example, the web page www.snopes.com/science/cricket.asp. Cited on page 406.

311 H. DE LANG, *Moleculaire gastronomie*, Nederlands tijdschrift voor natuurkunde 74, pp. 431–433, 2008. Cited on page 406.

312 EMILE BOREL, *Introduction géométrique à la physique*, Gauthier-Villars, 1912. Cited on page 406.

313 See V. L. TELEGDI, *Enrico Fermi in America*, Physics Today 55, pp. 38–43, June 2002. Cited on page 407.

314 K. SCHMIDT-NIELSEN, *Desert Animals: Physiological Problems of Heat and Water*, Oxford University Press, 1964. Cited on page 408.

315 Following a private communication by Richard Rusby, this is the value of 1997, whereas it was estimated as 99.975°C in 1989, as reported by GARETH JONES & RICHARD RUSBY, *Official: water boils at 99.975°C*, Physics World 2, pp. 23–24, September 1989, and R. L. RUSBY, *Ironing out the standard scale*, Nature 338, p. 1169, March 1989. For more details on temperature measurements, see Ref. 287. Cited on pages 408 and 549.

316 Why entropy is created when information is erased, but not when it is acquired, is explained in C. H. BENNETT & R. LANDAUER, *Fundamental Limits of Computation*, Scientific American 253:1, pp. 48–56, 1985. The conclusion: we should pay to throw the newspaper away, not to buy it. Cited on page 409.

317 See, for example, G. SWIFT, *Thermoacoustic engines and refrigerators*, Physics Today 48, pp. 22–28, July 1995. Cited on page 412.

318 Quoted in D. CAMPBELL, J. CRUTCHFIELD, J. FARMER & E. JEN, *Experimental mathematics: the role of computation in nonlinear science*, Communications of the Association of Computing Machinery 28, pp. 374–384, 1985. Cited on page 415.

319 For more about the shapes of snowflakes, see the famous book by W. A. BENTLEY & W. J. HUMPHREYS, *Snow Crystals*, Dover Publications, New York, 1962. This second printing of the original from 1931 shows a large part of the result of Bentley's lifelong passion, namely several thousand photographs of snowflakes. Cited on page 415.

320 K. SCHWENK, *Why snakes have forked tongues*, Science 263, pp. 1573–1577, 1994. Cited on page 418.

321 Human hands do not have five fingers in around 1 case out of 1000. How does nature ensure this constancy? The detailed mechanisms are not completely known yet. It is known, though, that a combination of spatial and temporal self-organization during cell proliferation and differentiation in the embryo is the key factor. In this self-regulating system, the GLI3 transcription factor plays an essential role. Cited on page 419.

322 E. MARTÍNEZ, C. PÉREZ-PENICHET, O. SOTOLONGO-COSTA, O. RAMOS, K. J. MÅLØY, S. DOUADY, E. ALTSHULER, *Uphill solitary waves in granular flows*, Physical Review 75, p. 031303, 2007, and E. ALTSHULER, O. RAMOS, E. MARTÍNEZ, A. J. BATISTA-LEYVA, A. RIVERA & K. E. BASSLER, *Sandpile formation by revolving rivers*, Physical Review Letters 91, p. 014501, 2003. Cited on page 420.

323 P. B. UMBANHOWAR, F. MELO & H. L. SWINNEY, *Localized excitations in a vertically vibrated granular layer*, Nature 382, pp. 793–796, 29 August 1996. Cited on page 421.

324 D. K. CAMPBELL, S. FLACH & Y. S. KIVSHAR, *Localizing energy through nonlinearity and discreteness*, Physics Today 57, pp. 43–49, January 2004. Cited on page 421.

325 B. ANDREOTTI, *The song of dunes as a wave-particle mode locking*, Physical Review Letters 92, p. 238001, 2004. Cited on page 421.

326 D. C. MAYS & B. A. FAYBISHENKO, *Washboards in unpaved highways as a complex dynamic system*, Complexity 5, pp. 51–60, 2000. See also N. TABERLET, S. W. MORRIS & J. N. MCELWAINE, *Washboard road: the dynamics of granular ripples formed by rolling wheels*, Physical Review Letters 99, p. 068003, 2007. Cited on pages 422 and 563.

327 K. KÖTTER, E. GOLES & M. MARKUS, *Shell structures with 'magic numbers' of spheres in a swirled disk*, Physical Review E 60, pp. 7182–7185, 1999. Cited on page 422.

328 A good introduction is the text by DANIEL WALGRAEF, *Spatiotemporal Pattern Formation, With Examples in Physics, Chemistry and Materials Science*, Springer 1996. Cited on page 422.

329 For an overview, see the Ph.D. thesis by JOCELINE LEGA, *Défauts topologiques associés à la brisure de l'invariance de translation dans le temps*, Université de Nice, 1989. Cited on page 424.

330 An idea of the fascinating mechanisms at the basis of the heart beat is given by A. BABLOYANTZ & A. DESTEXHE, *Is the normal heart a periodic oscillator?*, Biological Cybernetics 58, pp. 203–211, 1989. Cited on page 425.

331 For a short, modern overview of turbulence, see L. P. KADANOFF, *A model of turbulence*, Physics Today 48, pp. 11–13, September 1995. Cited on page 426.

332 For a clear introduction, see T. SCHMIDT & M. MAHRL, *A simple mathematical model of a dripping tap*, European Journal of Physics 18, pp. 377–383, 1997. Cited on page 426.

333 The mathematics of fur patterns has been studied in great detail. By varying parameters in reaction–diffusion equations, it is possible to explain the patterns on zebras, leopards, giraffes and many other animals. The equations can be checked by noting, for eyample, how the calculated patterns continue on the tail, which usually looks quite different. In fact, most patterns look differently if the fur is not flat but curved. This is a general phenomenon, valid also for the spot patterns of ladybugs, as shown by S. S. LIAW, C. C. YANG, R. T. LIU & J. T. HONG, *Turing model for the patterns of lady beetles*, Physical Review E 64, p. 041909, 2001. Cited on page 426.

334 An overview of science humour can be found in the famous anthology compiled by R. L. WEBER, edited by E. MENDOZA, *A Random Walk in Science*, Institute of Physics, 1973. It is also available in several expanded translations. Cited on page 426.

335 W. DREYBRODT, *Physik von Stalagmiten*, Physik in unserer Zeit pp. 25–30, Physik in unserer Zeit February 2009. Cited on page 427.

336 K. MERTENS, V. PUTKARADZE & P. VOROBIEFF, *Braiding patterns on an inclined plane*, Nature 430, p. 165, 2004. Cited on page 428.

337 These beautifully simple experiments were published in G. MÜLLER, *Starch columns: analog model for basalt columns*, Journal of Geophysical Research 103, pp. 15239–15253, 1998, in G. MÜLLER, *Experimental simulation of basalt columns*, Journal of Volcanology and Geothermal Research 86, pp. 93–96, 1998, and in G. MÜLLER, *Trocknungsrisse in Stärke*, Physikalische Blätter 55, pp. 35–37, 1999. Cited on page 429.

338 To get a feeling for viscosity, see the fascinating text by STEVEN VOGEL, *Life in Moving Fluids: the Physical Biology of Flow*, Princeton University Press, 1994. Cited on page 430.

339 B. HOF, C. W. H. VAN DOORNE, J. WESTERWEEL, F. T. M. NIEUWSTADT, H. WEDIN, R. KERSWELL, F. WALEFFE, H. FAISST & B. ECKHARDT, *Experimental observation of nonlinear traveling waves in turbulent pipe flow*, Science 305, pp. 1594–1598, 2004. See also B. HOF & al., *Finite lifetime of turbulence in shear flows*, Nature 443, p. 59, 2006. Cited on page 430.

340 A fascinating book on the topic is KENNETH LAWS & MARTHA SWOPE, *Physics and the Art of Dance: Understanding Movement*, Oxford University Press 2002. See also KENNETH LAWS & M. LOTT, *Resource Letter PoD-1: The Physics of Dance*, American Journal of Physics 81, pp. 7–13, 2013. Cited on page 430.

341 The fascinating variation of snow crystals is presented in C. MAGONO & C. W. LEE, *Meteorological classification of natural snow crystals*, Journal of the Faculty of Science, Hokkaido University Ser. VII, II, pp. 321–325, 1966, also online at the eprints.lib.hokudai.ac.jp website. Cited on page 431.

342 JOSEF H. REICHHOLF, *Eine kurze Naturgeschichte des letzten Jahrtausends*, Fischer Verlag, 2007. Cited on page 431.

343 See for example, E. F. Bunn, *Evolution and the second law of thermodynamics*, American Journal of Physics 77, pp. 922–925, 2009. Cited on page 432.

344 See the paper J. Maldacena, S. H. Shenker & D. Stanford, *A bound on chaos*, free preprint at www.arxiv.org/abs/1503.01409. The bound has not be put into question yet. Cited on page 432.

345 A good introduction of the physics of bird swarms is T. Feder, *Statistical physics is for the birds*, Physics Today 60, pp. 28–30, October 2007. Cited on page 433.

346 The Nagel-Schreckenberg model for vehicle traffic, for example, explains how simple fluctuations in traffic can lead to congestions. Cited on page 433.

347 J. J. Lissauer, *Chaotic motion in the solar system*, Reviews of Modern Physics 71, pp. 835–845, 1999. Cited on page 434.

348 See Jean-Paul Dumont, *Les écoles présocratiques*, Folio Essais, Gallimard, 1991, p. 426. Cited on page 435.

349 For information about the number π, and about some other mathematical constants, the website oldweb.cecm.sfu.ca/pi/pi.html provides the most extensive information and references. It also has a link to the many other sites on the topic, including the overview at mathworld.wolfram.com/Pi.html. Simple formulae for π are

$$\pi + 3 = \sum_{n=1}^{\infty} \frac{n\, 2^n}{\binom{2n}{n}} \tag{164}$$

or the beautiful formula discovered in 1996 by Bailey, Borwein and Plouffe

$$\pi = \sum_{n=0}^{\infty} \frac{1}{16^n} \left(\frac{4}{8n+1} - \frac{2}{8n+4} - \frac{1}{8n+5} - \frac{1}{8n+6} \right) . \tag{165}$$

The mentioned site also explains the newly discovered methods for calculating specific binary digits of π without having to calculate all the preceding ones. The known digits of π pass all tests of randomness, as the mathworld.wolfram.com/PiDigits.html website explains. However, this property, called *normality*, has never been proven; it is the biggest open question about π. It is possible that the theory of chaotic dynamics will lead to a solution of this puzzle in the coming years.

Another method to calculate π and other constants was discovered and published by D. V. Chudnovsky & G. V. Chudnovsky, *The computation of classical constants*, Proceedings of the National Academy of Sciences (USA) 86, pp. 8178–8182, 1989. The Chudnowsky brothers have built a supercomputer in Gregory's apartment for about 70 000 euros, and for many years held the record for calculating the largest number of digits of π. They have battled for decades with Kanada Yasumasa, who held the record in 2000, calculated on an industrial supercomputer. From 2009 on, the record number of (consecutive) digits of π has been calculated on a desktop PC. The first was Fabrice Bellard, who needed 123 days and used a Chudnovsky formula. Bellard calculated over 2.7 million million digits, as told on bellard.org. Only in 2019 did people start to use cloud computers. The present record is 31.415 million million digits. For the most recent records, see en.wikipedia.org/wiki/Chronology_of_computation_of_%CF%80. New formulae to calculate π are still occasionally discovered.

For the calculation of Euler's constant γ see also D. W. DeTemple, *A quicker convergence to Euler's constant*, The Mathematical Intelligencer, pp. 468–470, May 1993. Cited on pages 436 and 467.

350 The Johnson quote is found in WILLIAM SEWARD, *Biographiana*, 1799. For details, see the story in quoteinvestigator.com/2014/11/08/without-effort/. Cited on page 440.

351 The first written record of the letter U seems to be LEON BATTISTA ALBERTI, *Grammatica della lingua toscana*, 1442, the first grammar of a modern (non-latin) language, written by a genius that was intellectual, architect and the father of cryptology. The first written record of the letter J seems to be ANTONIO DE NEBRIJA, *Gramática castellana*, 1492. Before writing it, Nebrija lived for ten years in Italy, so that it is possible that the I/J distinction is of Italian origin as well. Nebrija was one of the most important Spanish scholars. Cited on page 442.

352 For more information about the letters thorn and eth, have a look at the extensive report to be found on the website www.everytype.com/standards/wynnyogh/thorn.html. Cited on page 442.

353 For a modern history of the English language, see DAVID CRYSTAL, *The Stories of English*, Allen Lane, 2004. Cited on page 442.

354 HANS JENSEN, *Die Schrift*, Berlin, 1969, translated into English as *Sign, Symbol and Script: an Account of Man's Efforts to Write*, Putnam's Sons, 1970. Cited on page 442.

355 DAVID R. LIDE, editor, *CRC Handbook of Chemistry and Physics*, 78th edition, CRC Press, 1997. This classic reference work appears in a new edition every year. The full Hebrew alphabet is given on page 2-90. The list of abbreviations of physical quantities for use in formulae approved by ISO, IUPAP and IUPAC can also be found there.
However, the ISO 31 standard, which defines these abbreviations, costs around a thousand euro, is not available on the internet, and therefore can safely be ignored, like any standard that is supposed to be used in teaching but is kept inaccessible to teachers. Cited on pages 444 and 446.

356 See the mighty text by PETER T. DANIELS & WILLIAM BRIGHT, *The World's Writing Systems*, Oxford University Press, 1996. Cited on page 445.

357 The story of the development of the numbers is told most interestingly by GEORGES IFRAH, *Histoire universelle des chiffres*, Seghers, 1981, which has been translated into several languages. He sums up the genealogy of the number signs in ten beautiful tables, one for each digit, at the end of the book. However, the book itself contains factual errors on every page, as explained for example in the review found at www.ams.org/notices/200201/rev-dauben.pdf and www.ams.org/notices/200202/rev-dauben.pdf. Cited on page 445.

358 See the for example the fascinating book by STEVEN B. SMITH, *The Great Mental Calculators – The Psychology, Methods and Lives of the Calculating Prodigies*, Columbia University Press, 1983. The book also presents the techniques that they use, and that anybody else can use to emulate them. Cited on page 446.

359 See for example the article 'Mathematical notation' in the *Encyclopedia of Mathematics*, 10 volumes, Kluwer Academic Publishers, 1988–1993. But first all, have a look at the informative and beautiful jeff560.tripod.com/mathsym.html website. The main source for all these results is the classic and extensive research by FLORIAN CAJORI, *A History of Mathematical Notations*, 2 volumes, The Open Court Publishing Co., 1928–1929. The square root sign is used in CHRISTOFF RUDOLFF, *Die Coss*, Vuolfius Cephaleus Joanni Jung: Argentorati, 1525. (The full title was *Behend vnnd Hubsch Rechnung durch die kunstreichen regeln Algebre so gemeinlicklich die Coss genent werden. Darinnen alles so treülich an tag gegeben, das auch allein auss vleissigem lesen on allen mündtlichē vnterricht mag begriffen werden, etc.*) Cited on page 446.

360 J. Tschichold, *Formenwamdlungen der et-Zeichen*, Stempel AG, 1953. Cited on page 448.

361 Malcolm B. Parkes, *Pause and Effect: An Introduction to the History of Punctuation in the West*, University of California Press, 1993. Cited on page 448.

362 This is explained by Berthold Louis Ullman, *Ancient Writing and its Influence*, 1932. Cited on page 448.

363 Paul Lehmann, *Erforschung des Mittelalters – Ausgewählte Abhandlungen und Aufsätze*, Anton Hiersemann, 1961, pp. 4–21. Cited on page 448.

364 Bernard Bischoff, *Paläographie des römischen Altertums und des abendländischen Mittelalters*, Erich Schmidt Verlag, 1979, pp. 215–219. Cited on page 448.

365 Hutton Webster, *Rest Days: A Study in Early Law and Morality*, MacMillan, 1916. The discovery of the unlucky day in Babylonia was made in 1869 by George Smith, who also rediscovered the famous *Epic of Gilgamesh*. Cited on page 449.

366 The connections between Greek roots and many French words – and thus many English ones – can be used to rapidly build up a vocabulary of ancient Greek without much study, as shown by the practical collection by J. Chaineux, *Quelques racines grecques*, Wetteren – De Meester, 1929. See also Donald M. Ayers, *English Words from Latin and Greek Elements*, University of Arizona Press, 1986. Cited on page 450.

367 In order to write well, read William Strunk & E. B. White, *The Elements of Style*, Macmillan, 1935, 1979, or Wolf Schneider, *Deutsch für Kenner – Die neue Stilkunde*, Gruner und Jahr, 1987. Cited on page 451.

368 *Le Système International d'Unités*, Bureau International des Poids et Mesures, Pavillon de Breteuil, Parc de Saint Cloud, 92310 Sèvres, France. All new developments concerning SI units are published in the journal *Metrologia*, edited by the same body. Showing the slow pace of an old institution, the BIPM launched a website only in 1998; it is now reachable at www.bipm.fr. See also the www.utc.fr/~tthomass/Themes/Unites/index.html website; this includes the biographies of people who gave their names to various units. The site of its British equivalent, www.npl.co.uk/npl/reference, is much better; it provides many details as well as the English-language version of the SI unit definitions. Cited on page 452.

369 The bible in the field of time measurement is the two-volume work by J. Vanier & C. Audoin, *The Quantum Physics of Atomic Frequency Standards*, Adam Hilge, 1989. A popular account is Tony Jones, *Splitting the Second*, Institute of Physics Publishing, 2000.

The site opdaf1.obspm.fr/www/lexique.html gives a glossary of terms used in the field. For precision *length* measurements, the tools of choice are special lasers, such as mode-locked lasers and frequency combs. There is a huge literature on these topics. Equally large is the literature on precision *electric current* measurements; there is a race going on for the best way to do this: counting charges or measuring magnetic forces. The issue is still open. On *mass* and atomic mass measurements, see the volume on relativity. On high-precision *temperature* measurements, see Ref. 287. Cited on page 453.

370 The unofficial SI prefixes were first proposed in the 1990s by Jeff K. Aronson of the University of Oxford, and might come into general usage in the future. See New Scientist 144, p. 81, 3 December 1994. Other, less serious proposals also exist. Cited on page 454.

371 The most precise clock built in 2004, a caesium fountain clock, had a precision of one part in 10^{15}. Higher precision has been predicted to be possible soon, among others by M. Takamoto, F.-L. Hong, R. Higashi & H. Katori, *An optical lattice clock*, Nature 435, pp. 321–324, 2005. Cited on page 456.

Vol. II, page 71

372 J. BERGQUIST, ed., *Proceedings of the Fifth Symposium on Frequency Standards and Metrology*, World Scientific, 1997. Cited on page 456.

373 J. SHORT, *Newton's apples fall from grace*, New Scientist 2098, p. 5, 6 September 1997. More details can be found in R. G. KEESING, *The history of Newton's apple tree*, Contemporary Physics **39**, pp. 377–391, 1998. Cited on page 457.

374 The various concepts are even the topic of a separate international standard, ISO 5725, with the title *Accuracy and precision of measurement methods and results*. A good introduction is JOHN R. TAYLOR, *An Introduction to Error Analysis: the Study of Uncertainties in Physical Measurements*, 2nd edition, University Science Books, Sausalito, 1997. Cited on page 458.

375 The most recent (2010) recommended values of the fundamental physical constants are found only on the website physics.nist.gov/cuu/Constants/index.html. This set of constants results from an international adjustment and is recommended for international use by the Committee on Data for Science and Technology (CODATA), a body in the International Council of Scientific Unions, which brings together the International Union of Pure and Applied Physics (IUPAP), the International Union of Pure and Applied Chemistry (IUPAC) and other organizations. The website of IUPAC is www.iupac.org. Cited on page 460.

376 Some of the stories can be found in the text by N. W. WISE, *The Values of Precision*, Princeton University Press, 1994. The field of high-precision measurements, from which the results on these pages stem, is a world on its own. A beautiful introduction to it is J. D. FAIRBANKS, B. S. DEAVER, C. W. EVERITT & P. F. MICHAELSON, eds., *Near Zero: Frontiers of Physics*, Freeman, 1988. Cited on page 460.

377 For details see the well-known astronomical reference, P. KENNETH SEIDELMANN, *Explanatory Supplement to the Astronomical Almanac*, 1992. Cited on page 465.

378 F.F. STANAWAY & al., *How fast does the Grim Reaper walk? Receiver operating characteristic curve analysis in healthy men aged 70 and over*, British Medical Journal 343, p. 7679, 2011. This paper by an Australian research team, was based on a study of 1800 older men that were followed over several years; the paper was part of the 2011 Christmas issue and is freely downloadable at www.bmj.com. Additional research shows that walking and training to walk rapidly can indeed push death further away, as summarized by K. JAHN & T. BRANDT, *Wie Alter und Krankheit den Gang verändern*, Akademie Aktuell 03, pp. 22–25, 2012, The paper also shows that humans walk upright since at least 3.6 million years and that walking speed decreases about 1 % per year after the age of 60. Cited on page 481.

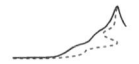

CREDITS

ACKNOWLEDGEMENTS

Many people who have kept their gift of curiosity alive have helped to make this project come true. Most of all, Peter Rudolph and Saverio Pascazio have been – present or not – a constant reference for this project. Fernand Mayné, Anna Koolen, Ata Masafumi, Roberto Crespi, Serge Pahaut, Luca Bombelli, Herman Elswijk, Marcel Krijn, Marc de Jong, Martin van der Mark, Kim Jalink, my parents Peter and Isabella Schiller, Mike van Wijk, Renate Georgi, Paul Tegelaar, Barbara and Edgar Augel, M. Jamil, Ron Murdock, Carol Pritchard, Richard Hoffman, Stephan Schiller, Franz Aichinger and, most of all, my wife Britta have all provided valuable advice and encouragement.

Many people have helped with the project and the collection of material. Most useful was the help of Mikael Johansson, Bruno Barberi Gnecco, Lothar Beyer, the numerous improvements by Bert Sierra, the detailed suggestions by Claudio Farinati, the many improvements by Eric Sheldon, the detailed suggestions by Andrew Young, the continuous help and advice of Jonatan Kelu, the corrections of Elmar Bartel, and in particular the extensive, passionate and conscientious help of Adrian Kubala.

Important material was provided by Bert Peeters, Anna Wierzbicka, William Beaty, Jim Carr, John Merrit, John Baez, Frank DiFilippo, Jonathan Scott, Jon Thaler, Luca Bombelli, Douglas Singleton, George McQuarry, Tilman Hausherr, Brian Oberquell, Peer Zalm, Martin van der Mark, Vladimir Surdin, Julia Simon, Antonio Fermani, Don Page, Stephen Haley, Peter Mayr, Allan Hayes, Norbert Dragon, Igor Ivanov, Doug Renselle, Wim de Muynck, Steve Carlip, Tom Bruce, Ryan Budney, Gary Ruben, Chris Hillman, Olivier Glassey, Jochen Greiner, squark, Martin Hardcastle, Mark Biggar, Pavel Kuzin, Douglas Brebner, Luciano Lombardi, Franco Bagnoli, Lukas Fabian Moser, Dejan Corovic, Paul Vannoni, John Haber, Saverio Pascazio, Klaus Finkenzeller, Leo Volin, Jeff Aronson, Roggie Boone, Lawrence Tuppen, Quentin David Jones, Arnaldo Uguzzoni, Frans van Nieuwpoort, Alan Mahoney, Britta Schiller, Petr Danecek, Ingo Thies, Vitaliy Solomatin, Carl Offner, Nuno Proença, Elena Colazingari, Paula Henderson, Daniel Darre, Wolfgang Rankl, John Heumann, Joseph Kiss, Martha Weiss, Antonio González, Antonio Martos, André Slabber, Ferdinand Bautista, Zoltán Gácsi, Pat Furrie, Michael Reppisch, Enrico Pasi, Thomas Köppe, Martin Rivas, Herman Beeksma, Tom Helmond, John Brandes, Vlad Tarko, Nadia Murillo, Ciprian Dobra, Romano Perini, Harald van Lintel, Andrea Conti, François Belfort, Dirk Van de Moortel, Heinrich Neumaier, Jarosław Królikowski, John Dahlman, Fathi Namouni, Paul Townsend, Sergei Emelin, Freeman Dyson, S.R. Madhu Rao, David Parks, Jürgen Janek, Daniel Huber, Alfons Buchmann, William Purves, Pietro Redondi, Damoon Saghian, Frank Sweetser, Markus Zecherle, Zach Joseph Espiritu, Marian Denes, Miles Mutka, plus a number of people who wanted to remain unnamed.

The software tools were refined with extensive help on fonts and typesetting by Michael Zedler and Achim Blumensath and with the repeated and valuable support of Donald Arseneau; help came also from Ulrike Fischer, Piet van Oostrum, Gerben Wierda, Klaus Böhncke, Craig Upright, Herbert Voss, Andrew Trevorrow, Danie Els, Heiko Oberdiek, Sebastian Rahtz, Don Story, Vincent Darley, Johan Linde, Joseph Hertzlinger, Rick Zaccone, John Warkentin, Ulrich Diez, Uwe Siart, Will Robertson, Joseph Wright, Enrico Gregorio, Rolf Niepraschk, Alexander Grahn, Werner Fabian and Karl Köller.

The typesetting and book design is due to the professional consulting of Ulrich Dirr. The typography was much improved with the help of Johannes Küster and his Minion Math font. The design of the book and its website also owe much to the suggestions and support of my wife Britta.

I also thank the lawmakers and the taxpayers in Germany, who, in contrast to most other countries in the world, allow residents to use the local university libraries.

From 2007 to 2011, the electronic edition and distribution of the Motion Mountain text was generously supported by the Klaus Tschira Foundation.

FILM CREDITS

The beautiful animations of the rotating attached dodecaeder on page 90 and of the embedded ball on page 168 are copyright and courtesy of Jason Hise; he made them for this text and for the Wikimedia Commons website. Several of his animations are found on his website www.entropygames.net. The clear animation of a suspended spinning top, shown on page 148, was made for this text by Lucas Barbosa. The impressive animation of the Solar System on page 155 was made for this text by Rhys Taylor and is now found at his website www.rhysy.net. The beautiful animation of the lunation on page 190 was calculated from actual astronomical data and is copyright and courtesy by Martin Elsässer. It can be found on his website www.mondatlas. de/lunation.html. The beautiful film of geostationary satellites on page 195 is copyright and courtesy by Michael Kunze and can be found on his beautiful site www.sky-in-motion.de/en. The beautiful animation of the planets and planetoids on page 220 is copyright and courtesy by Hans-Christian Greier. It can be found on his wonderful website www.parallax.at. The film of an oscillating quartz on page 291 is copyright and courtesy of Micro Crystal, part of the Swatch Group, found at www.microcrystal.com. The animation illustrating group and wave velocity on page 299 and the animation illustrating the molecular motion in a sound wave on page 311 are courtesy and copyright of the ISVR at the University of Southampton. The film of the rogue wave on page 323 is courtesy and copyright of Amin Chabchoub; details can be found at journals.aps.org/prx/abstract/10.1103/PhysRevX.2.011015. The films of solitons on page 317 and of dromions on page 329 are copyright and courtesy by Jarmo Hietarinta. They can be found on his website users.utu.fi/hietarin. The film of leapfrogging vortex rings on page 380 is copyright and courtesy by Lim Tee Tai. It can be found via his fluid dynamics website serve.me.nus.edu.sg. The film of the growing snowflake on page 422 is copyright and courtesy by Kenneth Libbrecht. It can be found on his website www.its.caltech.edu/~atomic/snowcrystals.

IMAGE CREDITS

The photograph of the east side of the Langtang Lirung peak in the Nepalese Himalayas, shown on the front cover, is courtesy and copyright by Kevin Hite and found on his blog thegettingthere. com. The lightning photograph on page 14 is courtesy and copyright by Harald Edens and found on the www.lightningsafety.noaa.gov/photos.htm and www.weather-photography.com websites. The motion illusion on page 18 is courtesy and copyright by Michael Bach and found on his website www.michaelbach.de/ot/mot_rotsnake/index.html. It is a variation of the illusion by Kitaoka Akiyoshi found on www.ritsumei.ac.jp/~akitaoka and used here with his permission. The figures on pages 20, 57 and 207 were made especially for this text and are copyright by Luca Gastaldi. The high speed photograph of a bouncing tennis ball on page 20 is courtesy and copyright by the International Tennis Federation, and were provided by Janet Page. The figure of Etna on pages 22 and 150 is copyright and courtesy of Marco Fulle and found on the wonderful website www.stromboli.net. The famous photograph of the Les Poulains and its lighthouse by Philip Plisson on page 23 is courtesy and copyright by Pechêurs d'Images; see the websites www.plisson.com and www.pecheurs-d-images.com. It is also found in Plisson's magnus opus *La Mer*, a stunning book of photographs of the sea. The picture on page 23 of Alexander Tsukanov jumping from one ultimate wheel to another is copyright and courtesy of the Moscow State Circus. The photograph of a deer on page 25 is copyright and courtesy of Tony Rodgers and taken from his website www.flickr.com/photos/moonm. The photographs of speed measurement devices on page 35 are courtesy and copyright of the Fachhochschule Koblenz, of Silva, of Tracer and of Wikimedia. The graph on page 38 is redrawn and translated from the wonderful book by HENK TENNEKES, *De wetten van de vliegkunst - Over stijgen, dalen, vliegen en zweven*, Aramith Uitgevers, 1993. The photographs of the ping-pong ball on page 40 and of the dripping water tap on page 355 are copyright and courtesy of Andrew Davidhazy and found on his website www.rit.edu/~andpph. The photograph of the bouncing water droplet on page 40 are copyright and courtesy of Max Groenendijk and found on the website www.lightmotif.nl. The photograph of the precision sundial on page 45 is copyright and courtesy of Stefan Pietrzik and found at commons.wikimedia.org/wiki/Image:Präzissions-Sonnenuhr_mit_Sommerwalze. jpg The other clock photographs in the figure are from public domain sources as indicated. The graph on the scaling of biological rhythms on page 47 is drawn by the author using data from the European Molecular Biology Organization found at www.nature.com/embor/journal/ v6/n1s/fig_tab/7400425_f3.html and Enrique Morgado. The drawing of the human ear on page page 50 and on page 325 are copyright of Northwestern University and courtesy of Tim Hain; it is found on his website www.dizziness-and-balance.com/disorders/bppv/otoliths.html. The illustrations of the vernier caliper and the micrometer screw on page 54 and 65 are copyright of Medien Werkstatt, courtesy of Stephan Bogusch, and taken from their instruction course found on their website www.medien-werkstatt.de. The photo of the tiger on page 54 is copyright of Naples zoo (in Florida, not in Italy), and courtesy of Tim Tetzlaff; see their website at www.napleszoo.com. The other length measurement devices on page 54 are courtesy and copyright of Keyence and Leica Geosystems, found at www.leica-geosystems.com. The curvimeter photograph on page 55 is copyright and courtesy of Frank Müller and found on the www.wikimedia.org website. The crystal photograph on the left of page 58 is copyright and courtesy of Stephan Wolfsried and found on the www.mindat.org website. The crystal photograph on the right of page 58 is courtesy of Tullio Bernabei, copyright of Arch. Speleoresearch & Films/La Venta and found on the www.laventa.it and www.naica.com.mx websites. The hollow Earth figure on pages 60 is courtesy of Helmut Diel and was drawn by Isolde Diel. The wonderful photographs on page 69, page 150, page 179, page 217, page 213 and page 506 are courtesy and copyright by Anthony Ayiomamitis; the story of the photographs is told on his beautiful website at www.perseus.gr. The anticrepuscular photograph on page 71 is courtesy and copyright by Peggy Peterson. The rope images on page 72 are copyright and courtesy of Jakob Bohr. The image of the tight knot on page 73 is courtesy and copyright by Piotr Pieranski. The firing caterpillar figure of page 79 is courtesy and copyright of Stanley Caveney. The photograph of an airbag sensor on page 86 is courtesy and copyright of Bosch; the accelerometer picture is courtesy and copyright of Rieker Electronics; the three drawings of the human ear are copyright of Northwestern University and courtesy of Tim Hain and found on his website www.dizziness-and-balance. com/disorders/bppv/otoliths.html. The photograph of Orion on page 87 is courtesy and copyright by Matthew Spinelli; it was also featured on antwrp.gsfc.nasa.gov/apod/ap030207.html. On page 88, the drawing of star sizes is courtesy and copyright Dave Jarvis. The photograph of Regulus and Mars on page 88 is courtesy and copyright of Jürgen Michelberger and found on www.jmichelberger.de. On page 91, the millipede photograph is courtesy and copyright of David Parks and found on his website www.mobot.org/mobot/madagascar/image.asp?relation=A71. The photograph of the gecko climbing the bus window on page 91 is courtesy and copyright of Marcel Berendsen, and found on his website www.flickr.com/photos/berendm. The photograph of the amoeba is courtesy and copyright of Antonio Guillén Oterino and is taken from his wonderful website *Proyecto Agua* at www.flickr.com/photos/microagua. The photograph of N. decemspinosa on page 91 is courtesy and copyright of Robert Full, and found on his website rjf9.biol.berkeley.edu/twiki/bin/view/PolyPEDAL/LabPhotographs. The photograph of P. ruralis on page 91 is courtesy and copyright of John Brackenbury, and part of his wonderful collection on the website www.sciencephoto.co.uk. The photograph of the rolling spider on page 91 is courtesy and copyright of Ingo Rechenberg and can be found at www.bionik.tu-berlin.de, while the photo of the child somersaulting is courtesy and copyright of Karva Javi, and can be found at www.flickr.com/photos/karvajavi. The photographs of flagellar motors on page 93 are copyright and courtesy by Wiley & Sons and found at emboj.embopress.org/content/30/14/2972. The two wonderful films about bacterial flagella on page 94 and on page 94 are copyright and courtesy of the Graduate School of Frontier Biosciences at Osaka University. The beautiful photograph of comet McNaught on page 95 is courtesy and copyright by its discoveror, Robert McNaught; it is taken from his website at www.mso.anu.edu.au/~rmn and is found also on antwrp.gsfc.nasa.gov/apod/ap070122.html. The sonoluminsecence picture on page 96 is courtesy and copyright of Detlef Lohse. The photograph of the standard kilogram on page 100 is courtesy and copyright by the Bureau International des Poids et Mesures (BIPM). On page 108, the photograph of Mendeleyev's balance is copyright of Thinktank Trust and courtesy of Jack Kirby; it can be found on the www.birminghamstories.co.uk website. The photograph of the laboratory balance is copyright and courtesy of Mettler-Toledo. The photograph of the cosmonaut mass measurement device is courtesy of NASA. On page 115, the photographs of the power meters are courtesy and copyright of SRAM, Laser Components and Wikimedia. The measured graph of the walking human on page 122 is courtesy and copyright of Ray McCoy. On page 130, the photograph of the stacked gyros is cortesy of Wikimedia. The photograph of the clock that needs no winding up is copyright Jaeger-LeCoultre and courtesy of Ralph Stieber. Its history and working are described in detail in a brochure available from the company. The company's website is www.Jaeger-LeCoultre.com. The photograph of the ship lift at Strépy-Thieux on page 132 is courtesy and copyright of Jean-Marie Hoornaert and found on Wikimedia Commons. The photograph of the Celtic wobble stone on page 133 is courtesy and copyright of Ed Keath and found on Wikimedia Commons. The photograph of the star trails on page 137 is courtesy and copyright of Robert Schwartz; it was featured on apod.nasa.gov/apod/ap120802.html. The photograph of Foucault's gyroscope on page 142 is courtesy and copyright of the museum of the CNAM, the Conservatoire National des Arts et Métiers in Paris, whose website is at www.arts-et-metiers.net. The photograph of the laser gyroscope on page 142 is courtesy and copyright of JAXA, the Japan Aerospace Exploration Agency, and found on their website at jda.jaxa.jp. On page 143, the three-dimensional model of the gyroscope is copyright and courtesy of Zach Joseph Espiritu. The drawing of the precision laser gyroscope on page 144 is courtesy of Thomas Klügel and copyright of the Bundesamt für Kartographie und Geodäsie. The photograph of the instrument is courtesy and copyright of Carl Zeiss. The machine is located at the Fundamentalstation Wettzell, and its website is found at www.wettzell.ifag.de. The illustration of plate tectonics on page 149 is from a film produced by NASA's HoloGlobe project and can be found on svs.gsfc.nasa.gov/cgi-bin/details.cgi?aid=1288 The graph of the temperature record on page 154 is copyright and courtesy Jean Jouzel and Science/AAAS. The photograph of a car driving through the snow on page 156 is copyright and courtesy by Neil Provo at neilprovo.com. The photographs of a crane fly and of a hevring fly with their halteres on page 160 are by Pinzo, found on Wikimedia Commons, and by Sean Mc-

Cann from his website ibycter.com. The MEMS photograph and graph is copyright and courtesy of ST Microelectronics. On page 166, Figure 123 is courtesy and copyright of the international Gemini project (Gemini Observatory/Association of Universities for Research in Astronomy) at www.ausgo.unsw.edu.au and www.gemini.edu; the photograph with the geostationary satellites is copyright and courtesy of Michael Kunze and can be found just before his equally beautiful film at www.sky-in-motion.de/de/zeitraffer_einzel.php?NR=12. The photograph of the Earth's shadow on page 167 is courtesy and copyright by Ian R. Rees and found on his website at weaknuclearforce.wordpress.com/2014/03/30/earths-shadow. The basilisk running over water, page 169 and on the back cover, is courtesy and copyright by the Belgian group TERRA vzw and found on their website www.terravzw.org. The water strider photograph on page 170 is courtesy and copyright by Charles Lewallen. The photograph of the water robot on page 170 is courtesy and copyright by the American Institute of Physics. The allometry graph about running speed in mammals is courtesy and copyright of José Iriarte-Díaz and of The Journal of Experimental Biology; it is reproduced and adapted with their permission from the original article, Ref. 132, found at jeb.biologists.org/content/205/18/2897. The illustration of the motion of Mars on page 175 is courtesy and copyright of Tunc Tezel. The photograph of the precision pendulum clock on page 181 is copyright of Erwin Sattler OHG,Sattler OHG, Erwin and courtesy of Ms. Stephanie Sattler-Rick; it can be found at the www.erwinsattler.de website. The figure on the triangulation of the meridian of Paris on page 184 is copyright and courtesy of Ken Alder and found on his website www.kenalder.com. The photographs of the home version of the Cavendish experiment on page 185 are courtesy and copyright by John Walker and found on his website www.fourmilab.ch/gravitation/foobar. The photographs of the precision Cavendish experiment on page 186 are courtesy and copyright of the Eöt-Wash Group at the University of Washington and found at www.npl.washington.edu/eotwash. The geoid of page 189 is courtesy and copyright by the GeoForschungsZentrum Potsdam, found at www.gfz-potsdam.de. The moon maps on page 191 are courtesy of the USGS Astrogeology Research Program, astrogeology.usgs.gov, in particular Mark Rosek and Trent Hare. The graph of orbits on page 193 is courtesy and copyright of Geoffrey Marcy. On page 196, the asteroid orbit is courtesy and copyrigt of Seppo Mikkola. The photograph of the tides on page 197 is copyright and courtesy of Gilles Régnier and found on his website www.gillesregnier.com; it also shows an animation of that tide over the whole day. The wonderful meteor shower photograph on page 206 is courtesy and copyright Brad Goldpaint and found on its website goldpaintphotography.com; it was also featured on apod.nasa.gov/apod/ap160808.html. The meteor photograph on page 206 is courtesy and copyright of Robert Mikaelyan and found on his website www.fotoarena.nl/tag/robert-mikaelyan/. The pictures of fast descents on snow on page 209 are copyright and courtesy of Simone Origone, www.simoneorigone.it, and of Éric Barone, www.ericbarone.com. The photograph of the Galilean satellites on page 210 is courtesy and copyright by Robin Scagell and taken from his website www.galaxypix.com. On page 217, the photographs of Venus are copyrigt of Wah! and courtesy of Wikimedia Commons; see also apod.nasa.gov/apod/ap060110.html. On page 218, the old drawing of Le Sage is courtesy of Wikimedia. The picture of the celestial bodies on page 222 is copyright and courtesy of Alex Cherney and was featured on apod.nasa.gov/apod/ap160816.html.

The pictures of solar eclipses on page 223 are courtesy and copyright by the Centre National d'Etudes Spatiales, at www.cnes.fr, and of Laurent Laveder, from his beautiful site at www.PixHeaven.net. The photograph of water parabolae on page 227 is copyright and courtesy of Oase GmbH and found on their site www.oase-livingwater.com. The photograph of insect gear on page 245 is copyright and courtesy of Malcolm Burrows; it is found on his website www.zoo.cam.ac.uk/departments/insect-neuro. The pictures of daisies on page 246 are copyright and courtesy of Giorgio Di Iorio, found on his website www.flickr.com/photos/gioischia, and of Thomas Lüthi, found on his website www.tiptom.ch/album/blumen/. The photograph of fireworks in Chantilly on page 249 is courtesy and copyright of Christophe Blanc and taken from his beautiful website at christopheblanc.free.fr. On page 258, the beautiful photograph of M74 is copyright and courtesy of Mike Hankey and found on his beautiful website cdn.mikesastrophotos.com. The figure of myosotis on page 267 is courtesy and copyright by Markku Savela. The image of the wallpaper groups on page page 268 is copyright and courtesy of Dror Bar-Natan, and is taken from his fascinating website at www.math.toronto.edu/~drorbn/Gallery. The images of solid symmetries on page page 269 is copyright and courtesy of Jonathan Goss, and is taken from his website at www.phys.ncl.ac.uk/staff.njpg/symmetry. Also David Mermin and Neil Ashcroft have given their blessing to the use. On page 292, the Fourier decomposition graph is courtesy Wikimedia. The drawings of a ringing bell on page 292 are courtesy and copyright of H. Spiess. The image of a vinyl record on page 293scale=1 is copyright of Chris Supranowitz and courtesy of the University of Rochester; it can be found on his expert website at www.optics.rochester.edu/workgroups/cml/opt307/spr05/chris. On page 296, the water wave photographs are courtesy and copyright of Eric Willis, Wikimedia and allyhook. The interference figures on page 303 are copyright and courtesy of Rüdiger Paschotta and found on his free laser encyclopedia at www.rp-photonics.com. On page 308, the drawings of the larynx are courtesy Wikimedia. The images of the microanemometer on page page 312 are copyright of Microflown and courtesy of Marcin Korbasiewicz. More images can be found on their website at www.microflown.com. The image of the portable ultrasound machine on page 313 is courtesy and copyright General Electric. The ultrasound image on page 313 courtesy and copyright Wikimedia. The figure of the soliton in the water canal on page 316 is copyright and courtesy of Dugald Duncan and taken from his website on www.ma.hw.ac.uk/solitons/soliton1.html. The photograph on page 319 is courtesy and copyright Andreas Hallerbach and found on his website www.donvanone.de. The image of Rubik's cube on page 346 is courtesy of Wikimedia. On page 327, the photographs of shock waves are copyright and courtesy of Andrew Davidhazy, Gary Settles and NASA. The photographs of wakes on page 328 are courtesy Wikimedia and courtesy and copyright of Christopher Thorn. On page 330, the photographs un unusual water waves are copyright and courtesy of Diane Henderson, Anonymous and Wikimedia. The fractal mountain on page 334 is copyright and courtesy by Paul Martz, who explains on his website www.gameprogrammer.com/fractal.html how to program such images. The photograph of the oil droplet on a water surface on page 335 is courtesy and copyright of Wolfgang Rueckner and found on sciencedemonstrations.fas.harvard.edu/icb. The soap bubble photograph on page page 342 is copyright and courtesy of LordV and found on his website www.flickr.com/photos/lordv. The photographs of silicon carbide on page 343 are copyright and courtesy of Dietmar Siche. The photograph of a single barium ion on page 344 is copyright and courtesy of Werner Neuhauser at the Universität Hamburg. The AFM image of silicon on page 344 is copyright of the Universität Augsburg and is used by kind permission of German Hammerl. The figure of helium atoms on metal on page 344 is copyright and courtesy of IBM. The photograph of an AFM on page 345 is copyright of Nanosurf (see www.nanosurf.ch) and used with kind permission of Robert Sum. The photograph of the tensegrity tower on page 349 is copyright and courtesy of Kenneth Snelson. The photograph of the Atomium on page 351 is courtesy and copyright by the Asbl Atomium Vzw and used with their permission, in cooperation with SABAM in Belgium. Both the picture and the Atomium itself are under copyright. The photographs of the granular jet on page 352 in sand are copyright and courtesy of Amy Shen, who discovered the phenomenon together with Sigurdur Thoroddsen. The photographs of the machines on page 352 are courtesy and copyright ASML and Voith. The photograph of the bucket-wheel excavator on page 353 is copyright and courtesy of RWE and can be found on their website www.rwe.com. The photographs of fluid motion on page 355 are copyright and courtesy of John Bush, Massachusetts Institute of Technology, and taken from his website www-math.mit.edu/~bush. On page 360, the images of the fluid paradoxa are courtesy and copyright of IFE. The images of the historic Magdeburg experiments by Guericke on page 361 are copyright of Deutsche Post, Otto-von-Guericke-Gesellschaft at www.ovgg.ovgu.de, and the Deutsche Fotothek at www.deutschefotothek.de; they are used with their respective permissions. On page page 362, the laminar flow photograph is copyright and courtesy of Martin Thum and found on his website at www.flickr.com/photos/39904644@N05; the melt water photograph is courtesy and copyright of Steve Butler and found on his website at www.flickr.com/photos/11665506@N00. The sailing boat on page 363 is courtesy and copyright of Bladerider International. The illustration of the atmosphere on page 365 is copyright of Sebman81 and courtesy of Wikimedia. The impressive computer image about the amount of water on Earth on page 371 is copyright and courtesy of Jack Cook, Adam Nieman, Woods Hole Oceanographic Institution, Howard Perlman and USGS; it was featured on apod.nasa.gov/apod/ap160911.html. The figures of wind speed measurement systems on page 375 are courtesy and copyright of AQSystems, at www.aqs.se, and Leosphere at www.leosphere.fr On page 377, the Leidenfrost photographs are courtesy and copyright Kenji Lopez-Alt and found on www.seriouseats.com/2010/08/how-to-boil-water-faster-simmer-temperatures.html. The photograph of the smoke ring at Etna on page 379 is courtesy and copyright by Daniela Szczepanski and found at her extensive websites www.vulkanarchiv.de and www.vulkane.net. On page 381, the photographs of rolling droplets are copyright and courtesy of David Quéré and taken from iusti.polytech.univ-mrs.fr/~aussillous/marbles.htm. The thermographic images of a braking bicycle on page 384 are copyright Klaus-Peter Möllmann and Michael Vollmer, Fachhochschule Brandenburg/Germany, and courtesy of Michael Vollmer and Frank Pinno. The image of page 384 is courtesy and copyright of ISTA. The images of thermometers on page 387 are courtesy and copyright Wikimedia, Ron Marcus, Braun GmbH, Universum, Wikimedia and Thermodevices. The balloon photograph on page 390 is copyright Johan de Jong and courtesy of the Dutch Balloon Register found at www.dutchballoonregister.nl.ballonregister, nederlands The pollen image on page 392 is from the Dartmouth College Electron Microscope Facility and courtesy Wikimedia. The scanning tunnelling microscope picture of gold on page 401 is courtesy of Sylvie Rousset and copyright by CNRS in France. The photograph of the Ranque–Hilsch vortex tube is courtesy and copyright Coolquip. The photographs and figure on page 419 are copyright and courtesy of Ernesto Altshuler, Claro Noda and coworkers, and found on their website www.complexperiments.net. The road corrugation photo is courtesy of David Mays and taken from his paper Ref. 326. The oscillon picture on page 421 is courtesy and copyright by Paul Umbanhowar. The drawing of swirled spheres on page 421 is courtesy and copyright by Karsten Kötter. The pendulum fractal on page 425 is courtesy and copyright by Paul Nylander and found on his website bugman123.com. The fluid flowing over an inclined plate on page 427 is courtesy and copyright by Vakhtang Putkaradze. The photograph of the Belousov-Zhabotinski reaction on page 428 is courtesy and copyright of Yamaguchi University and found on their picture gallery at www.sci.yamaguchi-u.ac.jp/sw/sw2006/. The photographs of starch columns on page 429 are copyright of Gerhard Müller (1940–2002), and are courtesy of Ingrid Hörnchen. The other photographs on the same page are courtesy and copyright of Raphael Kessler, from his websitewww.raphaelk.co.uk, of Bob Pohlad, from his websitewww.ferrum.edu/bpohlad, and of Cédric Hüsler. On page 431, the diagram about snow crystals is copyright and courtesy by Kenneth Libbrecht; see his website www.its.caltech.edu/~atomic/snowcrystals. The photograph of a swarm of starlings on page 432 is copyright and courtesy of Andrea Cavagna and Physics Today. The table of the alphabet on page 441 is copyright and courtesy of Matt Baker and found on his website at usefulcharts.com. The photograph of the bursting soap bubble on page 477 is copyright and courtesy by Peter Wienerroither and found on his website homepage.univie.ac.at/Peter.Wienerroither. The photograph of sunbeams on page 480 is copyright and courtesy by Fritz Bieri and Heinz Rieder and found on their website www.beatenbergbilder.ch. The drawing on page 485 is courtesy and copyright of Daniel Hawkins. The photograph of a slide rule on page 489 is courtesy and copyright of Jörn Lütjens, and found on his website www.joernluetjens.de. On page 493, the bicycle diagram is courtesy and copyright of Arend Schwab. On page 506, the sundial photograph is courtesy and copyright of Stefan Pietrzik. On page 510 the chimney photographs are copyright and courtesy of John Glaser and Frank Siebner. The photograph of the ventomobil on page 518 is courtesy and copyright Tobias Klaus. All drawings are copyright by Christoph Schiller. If you suspect that your copyright is not correctly given or obtained, this has not been done on purpose; please contact me in this case.

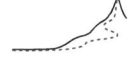

SUBJECT INDEX

B

BUREAU

C

CYRILLIC

E
——
EVOLUTION

G

GRAVITATIONAL

G

GRAVITATIONAL

K

KILOGRAM

M
—
METEOROID

N

N
‾‾
NUTATION

P

P

PRINCIPLE

S

S

STAINLESS

S

STAIRCASE

T

TOILET

W

WAVE

W

WAVELENGTH

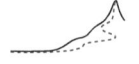

Printed in Great Britain
by Amazon

2697123gR00073